LIMS : IMPLEMENTATION AND MANAGEMENT

LIMS: Implementation and Management

Allen S. Nakagawa
Analytical Systems, Inc.
McDonald, Pennsylvania, USA

ROYAL
SOCIETY OF
CHEMISTRY

A catalogue record for this book is available from the British Library.

ISBN 0–85186–824–X

Published by The Royal Society of Chemistry,
Thomas Graham House, The Science Park, Cambridge CB4 4WF

Typeset by Vision Typesetting, Manchester
Printed and bound by Bookcraft (Bath) Ltd.

Preface

Each year millions of dollars are spent by corporations, universities, and government agencies worldwide on analysing samples for their physical, chemical, and biological properties. This requires substantial investment in facilities, equipment, reagents, instruments, and highly trained people.

This money is spent because testing information is needed to:

- determine the adequacy of existing processes and products;
- develop new products and innovative technologies;
- establish the presence or absence of undesirable conditions for workers, clients, or communities; and
- conform with reporting requirements or regulations.

Today every organization faces competitive pressures to control costs, accelerate innovations, and produce products or services of consistently superior quality. This forces everyone to seek new ways of operating more efficiently while simultaneously maximizing the strategic value of existing resources. Every laboratory and research manager is challenged by rapid advances in analytical technologies, a shortage of trained personnel, and increased pressures to rapidly generate, disseminate, and maintain results that favourably impact their organizations bottom-line.

Recent innovations in automated samplers, laboratory robotics, microprocessor-controlled instruments, and the refinement of innovative analytical techniques has dramatically increased laboratory data production and testing capabilities. These newer systems create so much data that scientists are forced to direct more and more attention to information management.

Laboratory Information Management Systems (LIMS) provide laboratory information handling functions such as sample tracking, data analysis, calculations, scheduling, data collation, reporting, *etc.* The available offerings cover a wide spectrum of capabilities and costs. Options range from small personal computers worth a few thousand dollars to room-sized mainframes costing millions.

Unfortunately there are hidden costs associated with each LIMS. In addition to the purchase of computer hardware and software, more money is needed for:

- development efforts to make the system fit needs that are specific to the laboratory and that are not a part of the delivered system;
- trained personnel to keep the system running;

- maintenance contracts; and
- expendable Supplies.

Frequently these additional costs exceed the initial system's purchase price!

Are these costs worth it? Do they make organizations stronger or do they just increase overheads? Are LIMS inappropriate for certain laboratories? How should the technology be evaluated and implemented? These are questions facing each manager considering a LIMS. Rational answers to these questions are elusive due to rapid innovations in instrumentation, testing, and computer technology.

New technologies offer every technical manager countless possibilities for meeting future challenges by improving efficiency, throughput, and data quality, and by adding new testing capabilities. Unfortunately technology provides only tools, not solutions. Ultimately the success or failure of laboratories will depend on how well each organization can make these tools work.

This book is intended for managers responsible for the implementation of a LIMS. It defines relevant management issues and offers a framework for each laboratory to define an implementation strategy relevant to their needs. This book pragmatically presents issues and considerations that are important to three different types of managers who are essential to success. Often these managers are separated by disciplinary as well as organizational boundaries. They include laboratory management whose organization will be affected by the planned system, Information Systems management who must deal with the technical aspects of installation and subsequent maintenance of the system, and the LIMS project manager who must balance schedules, budgets, and personnel, and satisfy conflicting needs. Each contributor may have different interests, objectives, and reporting accountabilities; but their collective contributions to the implementation effort are important. Difficulties often arise when there are breakdowns in communications and understanding between managers responsible for day to day operation of the laboratory and the technologists who are responsible for installation and maintenance.

An important consideration for any LIMS project includes the critical role of those that will ultimately use the system. Users need to be intimately involved in the planning and design of the system. They ultimately determine if the system will be a success. Unfortunately, the importance of this group and their impact on success is generally overlooked by the laboratory, information systems, and project managers.

This book highlights key elements of successful implementations including:

- understanding information flow in the laboratory targeted for a LIMS;
- determining how to apply LIMS technology in the targeted laboratory;
- establishing a plan for implementation; and
- executing the plan.

The book is divided into five sections:
(I) *Introduction to LIMS* provides an historical perspective on past LIMSs implementations and discusses the various components of laboratory

automation and information management.
(II) *Understanding Laboratory Information Flow* presents factors that determine why information handling is different for each laboratory. These differences ultimately impact the design and implementation of the planned LIMS. Information Engineering techniques are presented to trace laboratory sample and information flow.
(III) *Developing an Automation Strategy* presents factors to consider in the LIMS automation strategy. It considers the relevance and effect of technology on the laboratory's business.
(IV) *LIMS Implementation Planning* discusses options for implementation and key considerations that are critical to every LIMS implementation.
(V) *Implementation Pragmatics* presents elements needed for successful implementation and discusses unanticipated problems that surface during execution of the implementation plan.

Contents

CHAPTER 1

Historical Perspective

Above all, the principal product of any laboratory is information. This information advances the state of our commerce, welfare, and the technologies upon which our businesses and societies rely. The analytical measurement (or information generating) capabilities of laboratories are based upon increasingly complex testing systems. They consist of a coherent blend of components representing innovations from several disciplines. In fact, the specialized field of analytical chemistry consists of an ever-expanding collage of scientific areas, including organic chemistry, biochemistry, mathematics, inorganic chemistry, optics, biotechnology, microbiology, physics, electronics, statistics, and immunology.

During the past three decades, automation has exponentially increased the laboratory's data generation and information management capabilities. This chapter presents a historical perspective on computer-mediated improvements in laboratory capabilities, efficiency, and overall operations. The first section presents a chronological overview of innovations affecting the usage and application of computers within the laboratory. The second section discusses the historical development and selected case histories of LIMS.

1 History of Laboratory Computer Utilization

Progress in automation of the laboratory parallels technical and commercial developments in computers. The earliest mainframe computer systems of the 1960s were extremely expensive and required highly skilled professionals to keep them running. Initially, only the wealthiest organizations could afford them. Their use was highly restricted, tightly controlled, and had to be scheduled far in advance. By today's standards, the earlier machines were extremely primitive. Over the years, technological developments have significantly improved the capabilities of, performance of, and access to automation. At the same time, the cost of computers has dramatically decreased. The capability of the earlier million dollar systems is now readily available in pocket sized calculators costing less than 10 dollars. The extent and use of automation by laboratories has increased as the emerging technology has developed, stabilized, and become more affordable through large-scale production and marketing.

The use of computers in laboratories proceeded through the following stages.

1

- Initially, scientists created programs for their own use. Usage was limited to individuals who were highly computer literate and to organizations willing to invest the substantial time needed to create software.
- Commercial software packages emerged which provided increased access to automation. This extended the utilization of software to organizations who were unwilling to invest in substantial software development efforts.
- The costs of computer hardware decreased substantially due to advances in solid state electronics and the introduction of personal computers. With each passing year, more power was provided at lower absolute costs. This improved the affordability of automation and extended its acceptance by more and more organizations.
- Software advanced to provide easy to use and intuitive user interaction. Scientists no longer needed to be computer literate to capitalize on automation's capabilities. This facilitated the expanded acceptance of automated systems.

Computation

The earlier uses of computers involved tedious and time consuming calculations. Examples include kinetic, thermodynamic, and statistical computations. It replaced labour-intensive and tedious work previously completed manually or with the assistance of mechanical calculators. The computer provided higher levels of accuracy, reduced errors, and produced significant labour savings.

Instrument Automation

Advances in analytical instrumentation significantly improved the testing throughput of laboratories. These were enabled by advances in electronic controls and the integration of instruments with computerized data acquisition and control systems.

Electronic Instrument Control

The early instruments were inherently mechanical or optical devices. Data from each measurement was individually recorded on paper or film. The physical manipulations required, and the recording of all relevant data, consumed considerable effort.

The introduction of electronic controls in the 1960s substantially decreased the amount of manual labour required for analytical measurements. As an example, generating a sample's visible spectra previously involved the manual recording of many data points from a spectrophotometer. The analyst incrementally changed the wavelength setting of the instrument before recording the next data point. Once all the points were recorded, the spectra was obtained by manually plotting all the points on graph paper. With electronic controls, the analyst initiates the instruments wavelength scanning and the spectra is automatically plotted on an attached recording device, thereby eliminating the need for the manual recording and graphing of data.

Computerized Instrument Control and Data Acquisition

The advent of the integrated circuit and minicomputers in the 1970s facilitated automation of instrument data acquisition and control. Computers were then capable of controlling many aspects of the analysis. The data were automatically captured, stored, calculated, and processed by programs on the computer.

Within a few years several instruments with dedicated processors and programs were commercialized. Examples include automated systems dedicated to several types of chromatography and spectroscopy marketed by firms such as Beckman, Hewlett-Packard, and Perkin Elmer. In the late 1970s, almost every instrument sold included some form of built-in automation. Mass marketing of these systems substantially increased their accessibility and affordability.

Automation made possible the routine acceptance of several advanced testing techniques previously considered too time consuming for routine use. Many analytical measurements were not practical for high-volume testing without the instrument control, data acquisition and processing capabilities afforded by the combination of computers and instrumentation. Examples include FTIR (Fourier Transform Infrared Spectroscopy) and GC/MS (Gas Chromatography/Mass Spectrometry).

The routine use of automated sample handling devices permitted testing to occur without the constant attention of an analyst. This facilitated unattended operation and 24 hour utilization of many testing systems.

Capabilities provided by automated instruments have drastically improved the efficiency of analytical testing. One estimate is that advances over the 1978–1983 time period alone resulted in a five-fold reduction in the required time per test.[1]

Analytical Data and Information Management

Advances in instrument automation produced substantial improvements in analytical throughput and information flow. Each instrument was capable of analysing more samples and producing ever-increasing amounts of data per day. Each existed as autonomous and independent systems, with little or no connections between them. The collation and management of all the data from the wide variety of automated instruments became the new bottleneck in operations. The instruments produced data faster than laboratory staff could collate, calculate, interpret, report, and manage all the results.

Laboratory Information Management Systems

LIMS emerged to address needs for managing the totality of the laboratory's analytical testing data and information. They were initially developed by individual organizations during the late 1960s and emerged as commercial products in the early 1980s.

[1] S. Reber, *Am. Lab.*, 1983, **15:2**, 80.

In 1984, eight commercially available LIMS were available.[2] By 1991 over 50 companies claimed to offer LIMS Products.[3] One estimate put the 1991 worldwide LIMS market at 130 million US dollars, with over half the revenues from after market support and services.[4]

Instrument Data Systems

Specialized instrument data systems also emerged during this period. They included a combination of hardware and software for the capture, processing, and management of data from one or more instruments. These data systems worked with instruments from several vendors. Capabilities for the transfer of information to a LIMS were also included. Developments have been most notable in dedicated data systems for chromatography.

Structural Information Management

Many organizations, notably those involved with research and product development, needed to manage their collection of information on the properties of specific compounds of interest. In many cases, studies and tests were needlessly repeated because previously analytical data could not be found. Programs emerged to manage the organization's collective information base on specific substances. They organized data according to the compounds' unique three-dimensional structures. This prevented confusion between isomers, which are substances with the same molecular formula, but with different physical structures and properties.

Scientific Word Processing

The written dissemination of technical information involves a combination of text, equations, structural drawings, charts, graphs, spectra, and chromatograms. These elements are needed for scientific documentation such as publications, study reports, product specifications, test methods, manufacturing instructions, and procedures.

Technical documentation requires provisions for special Greek or non-English characters, mathematical symbols, and multi-level subscripts and superscripts. The initial word processing software was incapable of meeting the unique needs of scientists. They handled standard text only. The other elements had to be created manually.

In the late 1980s, commercial software emerged that permitted the integration of text, mathematical expressions, structural drawings, and graphics from other applications. This substantially reduced the effort and improved the speed of written technical communications.[5,6]

[2] P. King, *Anal. Instrum. Comput.*, 1991, **1:1**, 43.
[3] *Sci. Comput. Autom.*, 1991, **8:1**, 64.
[4] *Sci. Comput. Autom.*, 1992, **8:11**, 67.
[5] C. K. Gerson and R. A. Love, *Anal. Chem.*, 1987, **59**, 1031A.
[6] J. F. Barstow, D. del Rey, and J. S. Laufer, *Am. Lab.*, 1988, **20:7**, 82.

Robotics

In the early 1980s, industrial robots were adapted to the routine preparation of samples for analytical measurement. The robot presented an alternative to the time-consuming and potentially hazardous manual manipulations of samples and reagents by laboratory staff. The scope of robot use expanded to include wet chemistry procedures, instrument analysis, and communication with other laboratory systems. It was found that robots were suitable for high volume and well defined testing operations.

Data Analysis

Initially, all data analysis was performed by using software programs written by each laboratory. Commercial data analysis software emerged in the late 1970s. The offerings included programs for statistical, quality, and graphical data analysis. The variety and quantity of software packages grew as newer data analysis techniques emerged.

Modelling and Simulations

The use of modelling and simulation software became prominent in the 1980s. Practitioners included businesses committed to speeding up their product development cycles and improving their utilization of technical research resources. The software incorporated the knowledge from theoretical chemistry to predict the properties (such as spectra, reactivity, and receptor site interactions) of selected compounds.

Modelling and simulation software substantially reduced the organization's overall product development effort. This approach rapidly screened numerous compounds and only identified those that were worthy of more detailed study. Once identified, only the most promising substances were synthesized and subjected to further laboratory studies.

Information Retrieval

Every technical organization relies on the knowledge developed by scientists from other laboratories and other organizations. Public information is disseminated through numerous journals, periodicals, and other publications. Information content providers are organizations who read, summarize, index, and store the key contents of information residing in the public domain. On-line services are offered by businesses who provide remote access to these databases. Scientists utilize these services to secure knowledge on current and past developments on any defined topic of interest. These systems save the scientist considerable time in searching the literature to maintain an awareness of current developments and newer techniques.

2 LIMS Case Studies

Selected case studies of LIMS implementations are presented in this section. Included is a presentation of their laboratory environment, the technical approach taken, and the benefits realized.

Ralston Purina Company

The Ralston Purina Company (St. Louis, MO, USA) LIMS project was initiated in 1968. It covers 300 users at a central research facility and branch laboratories. They collectively perform 450 000 analyses per year with 600 different procedures. Components of the Ralston Purina system included a LIMS, an instrument data acquisition and control system, and an experimental data analysis system.

The LIMS software and the experimental data analysis systems reside on an IBM mainframe. The LIMS handles the entry, storage, manipulation, retrieval, and printing of data on samples, sample loads, and work schedules. The experimental data analysis system provides for statistical analysis and specialized report generation. The instrument data acquisition and control system is based on PDP-11 minicomputers. It supports the direct data acquisition from a variety of laboratory instruments.

Implementation of the systems required approximately 25 years of effort on software development. Estimated benefits of the system included projected labour savings of 15 to 20 years of effort per year.[7]

Mobil Research and Development Corporation

The Mobil Research and Development Corporation (Paulsboro, NJ, USA) LIMS was initiated in 1969. The analytical services laboratory supports annual workloads of 60 000 samples and 200 000 tests performed according to 800 procedures. Testing is performed by 49 technicians. The LIMS was implemented on a DEC-20 mainframe. It provides reports on sample status, laboratory workloads, backlogs, and analytical results. A data analysis package allows researchers to establish individual data bases to create custom plots and to analyse data trends. Data transfer from automated instruments occurs through software on PDP-11 minicomputers.

The automation effort has increased productivity by approximately 64% during the 1969–1983 period. The laboratory is handling a 25% workload increase with 15 fewer technicians.[8]

Calgon Water Management

The analytical laboratories at Calgon Water Management (Pittsburgh, PA, USA) initiated a LIMS feasibility study in 1982. The analytical services

---

[7] E. L. Schneider, *Anal. Chem.*, 1982, **33**, 277A.
[8] R. J. Kobrin, *Anal. Chem.*, 1982, **33**, 270A.

laboratory performed 200 000 tests on 40 000 samples per year. Implementation of the system commenced in 1984 based on hardware and software from a commercial LIMS supplier. Customized extensions to the commercial software were added to accommodate the laboratory's work flow and management reporting. Major functions provided by the system included sample logging, sample status tracking, test scheduling, analytical data entry, calculations, results approval, sample reporting, partial results reporting, workload monitoring, and turnaround reports.

Implementation of the initially targeted functions was completed in late 1986, an elapsed time of 18 months. The effort involved a task force consisting of laboratory staff with the assistance of corporate Management Information Systems and the vendor. The laboratory experienced an overall productivity improvement of 30%, with payback of the system occurring within 1 year.[9]

Construction Technology Laboratories, Inc.

The Construction Technology Laboratories, Inc. (Skokie, IL, USA) provides cement-related testing services to the construction industry. It also provides contract research and development work on new products and processes. Much of the testing supports litigation. The laboratory performs 20 000 determinations on 5000 samples each year.

In 1988 the first attempt at automation employed a Macintosh spreadsheet program to track sample flow through the laboratory. It allowed the supervisor to monitor samples in the laboratory and the tests to be performed on each sample. However, the spreadsheet did not allow for scheduling and status tracking.

The initial system was replaced with a series of Macintosh-based databases for project management, sample logging, sample label printing, and invoicing. However, significant time was still required for supervisors to schedule tests and assign work.

In 1990, commercial LIMS software based on IBM personal computers was installed. This system provided for project management, sample logging, sample label printing, worklist creation, results entry, results reporting, project status reports, and invoicing. In 1991 two-thirds of the paperwork previously associated with testing was partially automated by the LIMS. Management estimated that the system annually saved 500 technician-hours for sample management.[10]

Eastman Kodak Company

Eastman Kodak (Rochester, NY, USA) embarked on a worldwide effort to facilitate information sharing among several laboratory sites in the USA, UK, France, and Mexico. Their strategy was first reported by Alan Uthman in 1990.[11]

[9] D. O. Bassett, *Am. Lab.*, 1987, **19:9**, 28.
[10] H. Kanare, *Sci. Comput. Autom.*, 1991, **7:11**, 33.
[11] A. P. Uthman, 'Development of an Analytical Information Management Strategy', Fourth International LIMS Conference, Pittsburgh, PA, USA, 1990.

It was based on:

- development and consolidation of data models for testing operations and worldwide manufacturing;
- use of a single LIMS software application platform;
- implementation of common information system practices and methodologies;
- implementation of a formal software quality assurance program;
- establishment of consistent operations between laboratories.

In 1992, four sites were implemented, eight were in process, and another four were planned during the next year. Their approach relied on a commercial LIMS product integrated with internally developed programs for chemical structure data management and report writing. Other software will be developed to handle functions not adequately addressed by the commercial LIMS software.

Benefits of their LIMS implementations include cycle time reductions, operation efficiency, error reduction, and enhanced services to manufacturing and customers.[12]

Miles, Inc.

The Miles, Inc. Polyurethane Quality Assurance Department (New Martinsville, WV, USA) laboratories provide on-site manufacturing support of commodity and specialty chemicals. They maintain continuous operation 24 hours per day, 365 days per year. They are staffed by 40 technicians for routine analysis with seven chemists for supervision and non-routine support. They report 288 000 results on 84 000 samples each year. Sample analysis includes raw materials, process control, and finished products. Analytical turnaround is expected in minutes and hours.

The project started in 1984 with creation of specifications. The system is based on commercial LIMS software with three separate databases. Two employees are assigned to LIMS maintenance and development. Other software is used for data charting and statistics. In 1992, finished product and process control sample testing was completed. Installation of raw material data was in progress.

Within the organization, the benefits and costs of the LIMS has had mixed reviews. The principal benefit of the system is its ability to facilitate documentation of laboratory quality control. This provides higher levels of customer satisfaction and confidence in the credibility of laboratory results and practices.[13]

[12] A. P. Uthman, *Chemometrics Intelligent Lab. Syst.: Lab. Inf. Manage.*, 1992, **17:3**, 295.
[13] T. J. Conti, *Chemometrics Intelligent Lab. Syst.: Lab. Inf. Manage.*, 1992, **17:2**, 301.

CHAPTER 2

What is a LIMS?

The exact definition of a LIMS is highly subjective. If you were to ask a group of people to detail what a LIMS consists of, many divergent and conflicting definitions would emerge. Perceptions differ between different organizations, between laboratories within a single organization, and even between individuals within the same laboratory group. The lack of a standard definition is particularly troublesome for those responsible for its implementation. A LIMS will fail if the initial expectations of those responsible for its use and funding do not coincide with what is ultimately delivered. The divergence in expectations increases with the number of people and groups involved.

This chapter presents various proposed definitions of a LIMS based on hardware, software, and functional criteria. Independent of current definitions or standards, a critical aspect of successful implementation is for you and your organization to consolidate and solidify your expectations for the LIMS. The discussion that follows presents the various dimensions of LIMS to hopefully help you in this process. Subsequent chapters present ways of understanding your own laboratory (Chapter 3), its role and internal interactions (Chapter 4), and its relationship to groups within and external to your organization (Chapter 5).

McDowall and Mattes[1] presented the following observations on the variety of LIMS definitions and their effects.

'Current LIMS definitions give us a vision of the accomplishments and the benefits without the understanding and insight necessary to implement a system. This lack of definition, combined with the variations in actual laboratory operation, results in LIMS meaning different things to different people.'

'Many organizations that have a limited view of LIMS have missed opportunities for significant scientific and business benefits'

1 Perceptions of a LIMS

This section discusses factors contributing to variations in the ways a LIMS can be perceived. It presents factors that result in differences between organizations and even between people within the same organization.

[1] R. D. McDowall and D. C. Mattes, *Anal. Chem.*, 1990, **62**, 1069A.

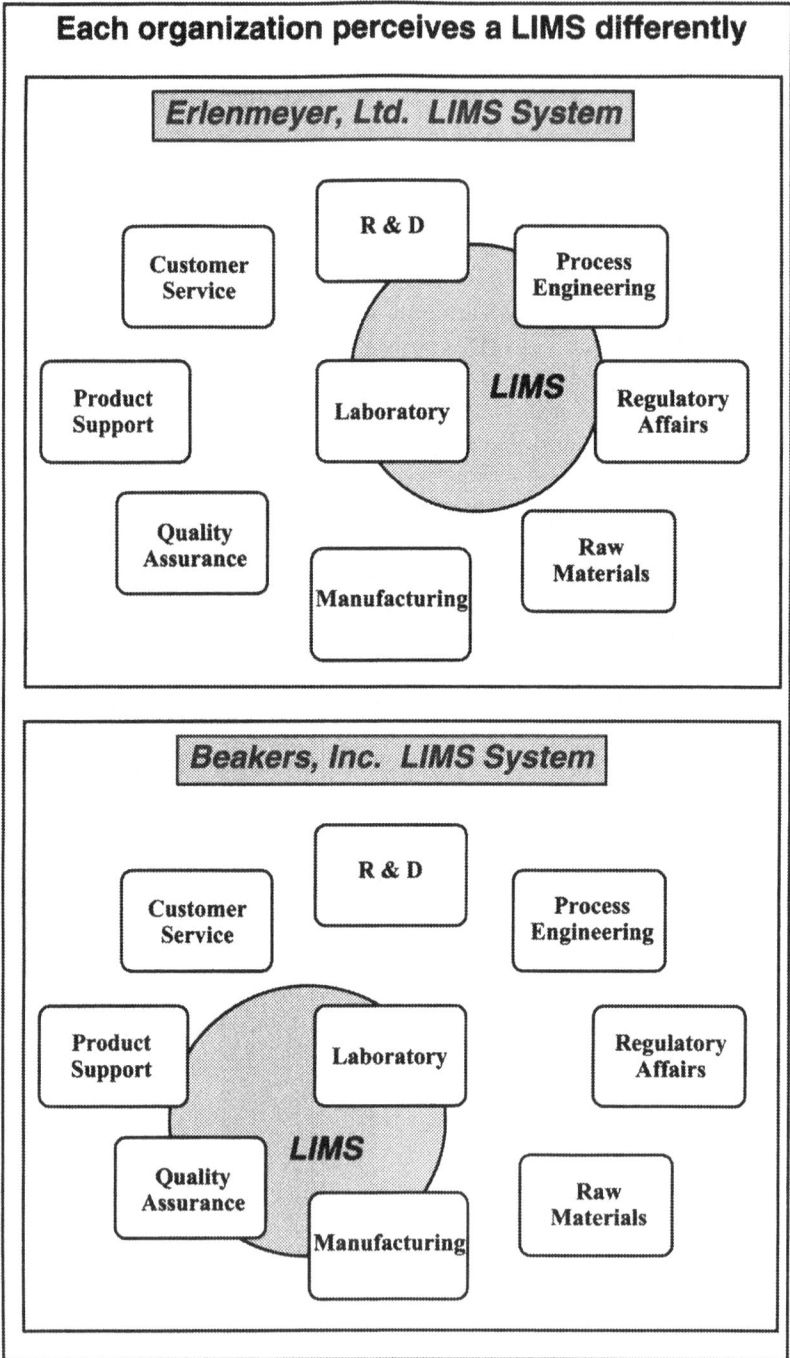

Figure 2-1 *Organizational Perceptions of a LIMS*

Differences Between Organizations

Each organization differs in its products, services, competitive position, and regulatory status. These factors contribute to the services provided and constraints placed upon their laboratories. Even within the same organization, there are typically several different types of laboratories. Some are involved in manufacturing support, others in customer service, and still others in research and development. Laboratories differ considerably based on the types of analysis performed, the kinds of samples analysed, and the ultimate end use of the test results. The organization, procedures, personnel, equipment, and technology utilized varies considerably between laboratories. Figure 2-1 contrasts the scope of a LIMS as approached by two different firms.

The information handled by each laboratory is dictated by what the laboratory generates and what their clients need. Information handled by a laboratory is also greatly affected by the underlying scientific discipline(s) supported by the laboratory. For example, information related to microscopic analysis differs considerably from chromatography. Microscopic data includes information regarding the specimen, the actual image of the specimen under the microscope, and conclusions based on an educated examination of morphological or other aspects of the image. However, the greatest portion of chromatography data are two-dimensional numerical recordings of detector response *versus* time. Various mathematical operations are performed on this data to identify components of the samples and their respective concentration values.

Needs for the management of laboratory information are dictated by the actual functions provided by the laboratory to support overall enterprise and client objectives. Rigorous information management protocols are needed if test results support critical decision-making processes, especially in highly regulated industries or in cases involving human health and welfare. Other factors driving information management needs include the volume of information within a laboratory, staffing constraints, anticipated workload projections, and turnaround times for sample processing.

Every system consists of a group of functionally related and interacting components supporting a common set of objectives. A LIMS system is composed of manual as well as automated components. The automated aspects include an impressive assortment of increasingly sophisticated computerized hardware and software technologies. However, the most sophisticated technology cannot provide solutions if they do not work in concert with manual business processes. The non-automated components include various organizational, procedural, personnel, and management aspects of the system.

Differences Within Organizations

Even within a given organization, different individuals will have varied perspectives regarding the definition of a LIMS. It means different things to different people. Too much emphasis on any one perspective can constrain the potential benefits that the system can deliver to the organization. A LIMS is somewhat like the proverbial elephant and three blind men. In the absence of sight, the men rely on their sense of touch. Each feels a different part of the creature's body and imagines it as being something different.

Business Executive Perspective

To a business executive, a LIMS provides a means of improving the handling and dissemination of laboratory results to business units outside the laboratory so that they can act on them. It is a way of moving the organization forward by improvements in customer service, product quality, operational efficiency, and development cycles.

Information Systems Perspective

To an Information Systems manager, a LIMS is another database application accompanied by hardware and specialized software for laboratories. It is but one of many applications within the organization that involves a myriad of sophisticated technologies. He or she is concerned about how the LIMS fits in with the other systems and applications.

Laboratory Perspectives

Laboratory Manager. To a laboratory manager concerned about handling increasing workloads, a LIMS provides a means of monitoring workload so that personnel can be moved to the laboratory groups with the highest demands. It also monitors the analytical turnaround and services provided to laboratory clients.

Laboratory Supervisor. To a laboratory group leader or supervisor, a LIMS helps track the status of samples and testing and provides information about work that has already been completed as well as work that is outstanding within their group. It assists in responding to client enquiries regarding the work performed on their behalf.

Laboratory Analyst. To a laboratory analyst, a LIMS provides a way of relieving the burden of repetitive calculations, processing, and transcriptions. It assists with, or totally eliminates, the handling of paperwork. It also provides them with tools to more effectively and efficiently plan their work.

2 Functional Model of a LIMS

In functional models, the LIMS is presented as several processes, each interacting with a central LIMS database. Figure 2-2 summarizes LIMS functions previously described by McDowall,[2] Mahaffey,[3] Scott,[4] and McDowall and Mattes.[1]

---

[2] R. D. McDowall, in 'Laboratory Information Management Systems — Concepts, Integration and Implementation', ed. R. D. McDowall, Sigma Press, Wilmslow, Cheshire, 1987, p. 1.
[3] R. R. Mahaffey, 'LIMS Applied Information Technology for the Laboratory', Van Nostrand Reinhold, New York, 1990.
[4] F. I. Scott, *Am. Lab.*, 1987, **19:11**, 50.

McDowall (1987)		**Scott (1987)**

Sample Management
Data Entry
Analysis Management
Laboratory Management
Reporting
Ad hoc Enquiries
Data Acquisition
Workload Management

Sample Log in
Sampling
Testing
Test Result Validation
Approval of Samples
Standard Report Generation
Ad hoc Reports
Updating
Archiving
Dictionaries

Mahaffey (1990)

McDowall & Mattes (1990)

Sample Log in
Collection Lists
Sample Receive
Worksheets
Worklists
Manual Results Entry
Automated Results Entry
Results Verification
Reporting Results

Data Capture
Data Analysis
Reporting
Management

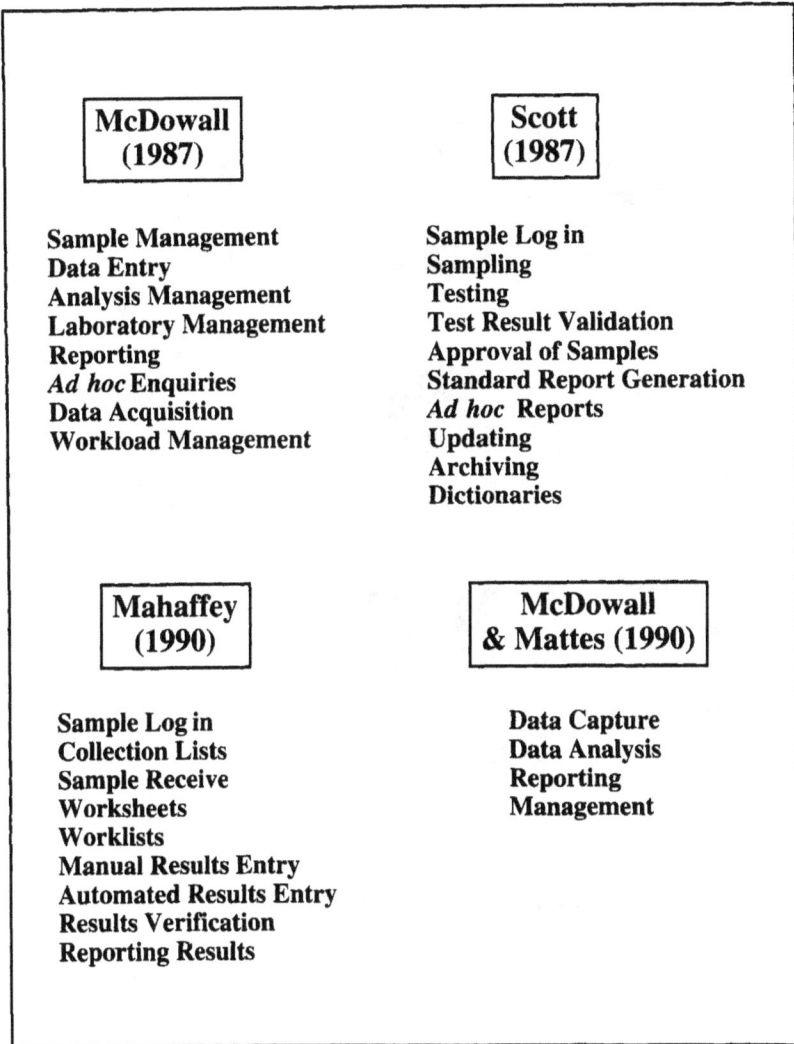

Figure 2-2 *LIMS Functions*

The remainder of this section is based on the LIMS model of Figure 2-3. Different people within the laboratory are affected by each component and its relative importance (and benefit) is perceived differently. Each part may be comprised of one or more supporting programs, entry screens, or reports. Conversely, a single program may incorporate several different functions. The quantity and sophistication of software elements for each function differs for each LIMS implementation.

A summarized description of each component follows. Other functions may apply to your laboratory. The LIMS can bring together several needs of the

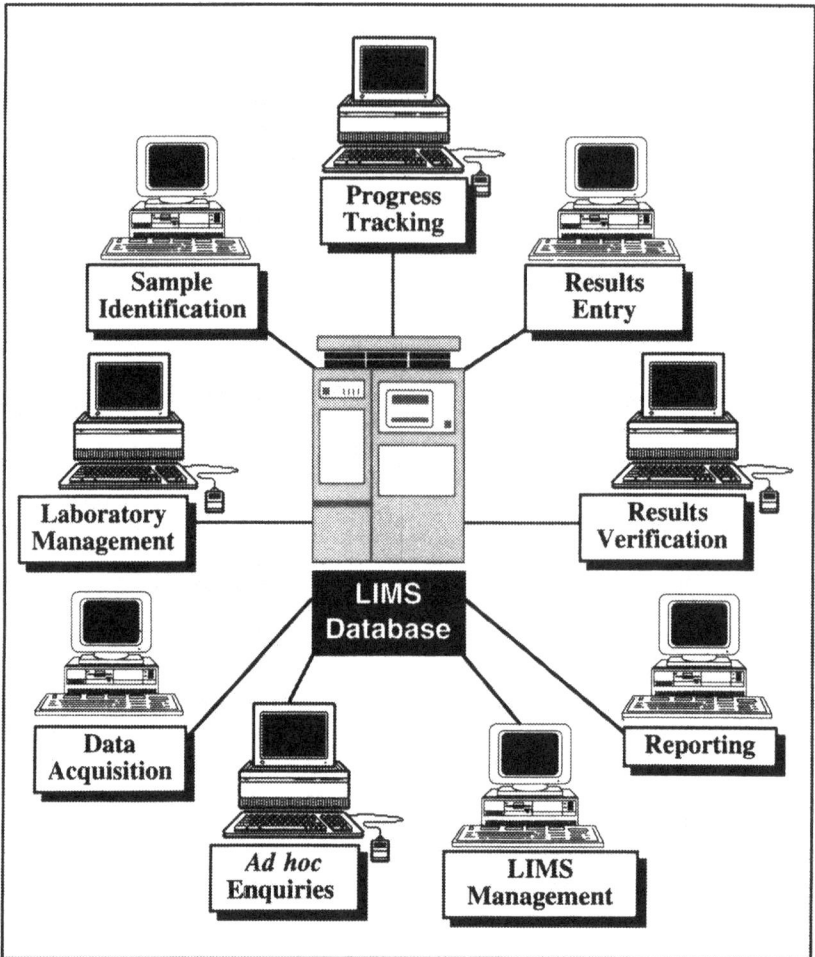

Figure 2-3 *Functional model of a LIMS*

organization. It can play an important role in the integration of many business functions outside the laboratory as shown in Figure 2-4.

Sample Identification

Sample Labelling

Samples are uniquely identified by the assignment of a distinctive identifier and the creation of labels which are physically attached to each sample container. The sample identifiers may be printed in a machine readable bar code as well as a human readable textual format. This facilitates the proper identification of samples as work is performed by other LIMS functions.

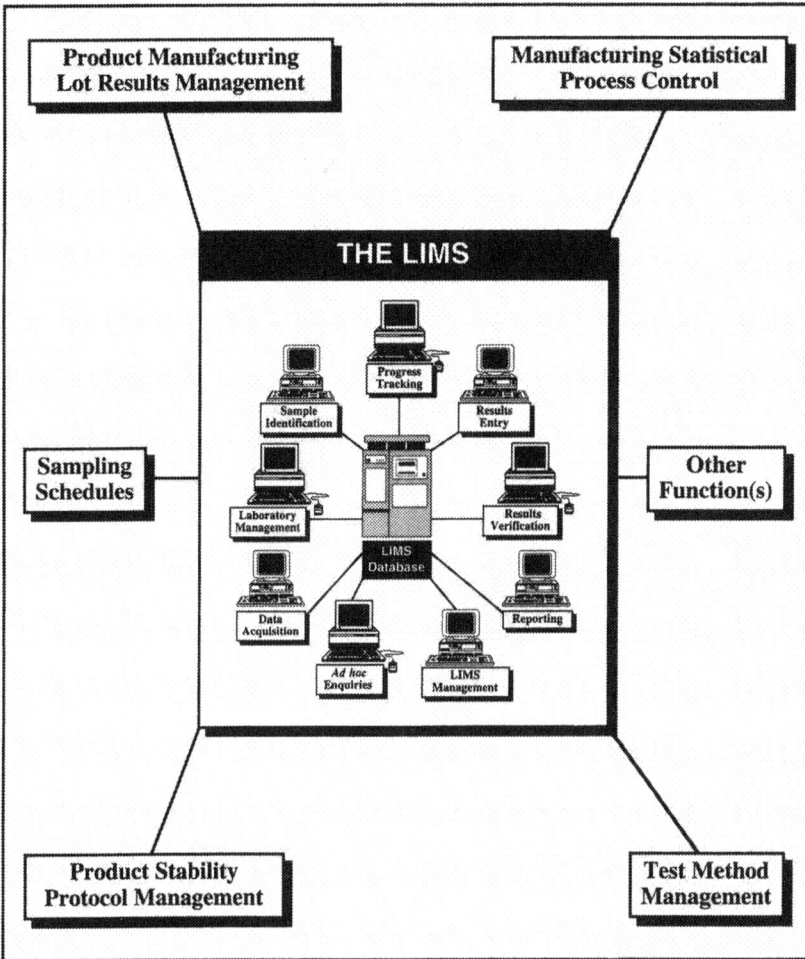

Figure 2-4 *LIMS integrated with other functions*

Sample Registration

Information about each sample is recorded. This includes descriptive information such as the sample type, sample source, sampling date, submitter, priority, and material type. The information may be manually transcribed from paperwork accompanying the sample, electronically transferred from another computer, or duplicated from testing programs already on the system.

Sample Log In

The date and time the sample is received in the laboratory is recorded. The related functions of sample registration and sample identification may occur before, at the same time, or following receipt in the laboratory.

Progress Tracking

Work Progress Tracking

The progress of assigned analytical work is tracked. This involves programs that update status states as work proceeds through the laboratory. Typical status states include logged in, complete, cancelled, and approved. The nomenclature and actual status states differ from one laboratory to another. Programs can also display the current progress of work in the laboratory.

Sample Location Tracking

The physical movement of samples from one laboratory location to another is tracked. This is important for cases in which a rigorous sample chain of custody must be maintained.

Results Creation

Test Data Entry

Data and test results arising from the analysis of samples is accepted and recorded in the database.

Test Calculations

Test data calculations previously defined within a method or protocol are performed. The outcome of those calculations are recorded in the LIMS database.

Results Verification

Analysis Review

Completed test results and supporting data are provided for review. If warranted, other functions are triggered by this review. This may include the correction of values entered in error or the initiation of retesting and resampling.

Analysis Approval

Authorized individuals release completed results from the laboratory.

Specification Checking

Result values are checked against previously established ranges of acceptability. In some cases a result may be evaluated against several specifications.

Analysis Modification

The analytical work currently assigned is modified. This may involve the assignment of additional tests, cancellation of tests, repeat testing, or resampling.

Results Modification

Individuals with the proper authorization are permitted to rectify incorrectly entered result values. Many regulated laboratories require an integrated audit trail to record the time, responsible individual, and reason for each modification. All originally recorded results and any subsequent modifications are stored in the LIMS database.

Reporting

Results Reporting

The analytical results obtained from testing a sample or series of samples is obtained from the LIMS database, collated, and formatted.

Results Distribution

Approved analytical results are electronically forwarded to designated parties. Results may also be transferred to other databases or programs for historical storage or further processing.

Report Definition

The information content and format of recurring reports are established. Once defined, the reports can be repeatedly created in several ways, on demand by authorized individuals or at regularly defined time intervals.

Ad hoc Enquiries

Ad hoc Query

Selected information is extracted from the database through a free form query. The query results may be displayed on a screen, printed as a report, or provided as a computer file for further manipulation.

Instrument Integration

Instrument Interfacing

The physical connection and capability for data transfer between an instrument and the LIMS is established. This involves matching of the data transfer protocols of both the instrument and the LIMS.

Instrument interfacing is often misunderstood. It only deals with issues related to the compatibility of physical connections and data transfer protocols between the instrument and the LIMS. Other functions are needed to actually transfer, process, and add data to the LIMS database. Interfacing an instrument does not ensure that it is integrated with the LIMS. However, none of the other instrument integration functions are possible without first establishing a proper interface.

Data Acquisition

The conversion of an instruments signal or data stream into a format that can be recognized and stored on a computer. It requires specialized hardware devices such as Analogue to Digital Converters (ADC) or Binary-Coded Decimal (BCD) Converters. Special software on the computer permits the capture and storage of data from these devices.

Instrument Control

The actual control of an instruments settings and operations is through instructions from a computer. Most instrument control functions are handled by computers which are a part of the instrument's integrated data system.

Run Lists

A run list presents candidate sample preparations, standards, and blanks to be sequentially run by an instrument. The actual sequence is determined by the analyst or by programs within the instrument.

Instrument Data Transfer

This involves the movement of data from the instrument or its data system to the LIMS. What is transferred may or may not be directly stored in the database. It may reside outside the database for further data reduction and processing before being added to the LIMS database.

Instrument Data Reduction

The transferred instrument data is calculated, reviewed, and further processed prior to storage in the LIMS database. This may involve the collation of data from several sources; other instruments as well as manually entered analytical results.

Laboratory Management

Work Scheduling

Work that is outstanding in the laboratory is selectively listed. It is used for the assignment and distribution of work within the laboratory. This can be done in several ways such as by analyst, work group, instrument, or test. This may involve the creation of work lists and worksheets.

Completed Analysis

Testing completed by the laboratory, or groups within the laboratory, is listed. It includes all testing completed during a defined time period. In some cases, the laboratory's service cycle time is also reported for all completed work. It details the analytical turnaround for various areas in the laboratory. Such a listing provides a focus for testing activities or groups in which performance is slipping and further investigation is required. It also highlights the areas in which improvements have been made.

Incomplete Work

An assessment of the total work backlog for a laboratory or laboratory group is given. It provides a way of determining if reallocation of personnel or overtime is required. It also provides a means of anticipating near term improvements or slippage in analytical turnaround time.

Workload Demands

All new work input to the laboratory during a defined time period is listed. Continuous workload demand monitoring provides the laboratory with the historical trend patterns for planning future personnel and equipment needs.

Protocol Maintenance

A repository is provided for the maintenance of active test methods, specifications, projects, studies, or other laboratory-specific protocols. It is referenced by other LIMS functions. Maintenance of a centralized repository ensures that protocols are executed consistently within the laboratory.

Pricing and Invoicing

The basis upon which customers are charged and costs allocated is maintained. Pricing can be based on tests, samples, or projects. Discounts may also be based on the quantity of services requested or on a per-customer basis. Uniform pricing policies are maintained on the system and applied whenever invoices are issued or costs are allocated.

Where necessary, invoices are issued for analytical services. They itemize the services provided, associated costs, and terms of payment. The invoicing function may be linked to an accounts payable function to track overdue invoices and customer payment histories.

Cost Allocation

The costs of analytical services during a defined time period are distributed among various laboratory clients. This includes services that are billed externally as well as those provided to other groups within the organization.

Systems Management

Security

Various functions are provided to protect the integrity of data and to maximize systems reliability. Included are functions such as:

- backup of the software and data;
- recovery from utility or other failures; and
- restricted access to data and programs.

Data Archive

Information that does not need to be immediately accessible and which is not frequently used is selectively copied, stored off-line, and removed from the active system. This optimizes the utilization of the systems storage resources and improves its overall response time.

Data Retrieval

When necessary, data from the historical archives are selectively restored on the active system in response to investigations or audits. Appropriate indexing and organization of the archives is necessary to ensure that data retrieval can be accomplished within a reasonable time period.

3 Architectural Model of a LIMS

Hardware and Software Architecture

Components of a LIMS include both computer hardware and software. Their relative relationships are shown in Figure 2-5. This section presents a general overview of the various hardware, software, and networking components of LIMS technology.

Hardware Components

The system's physical hardware components include its computer processor,

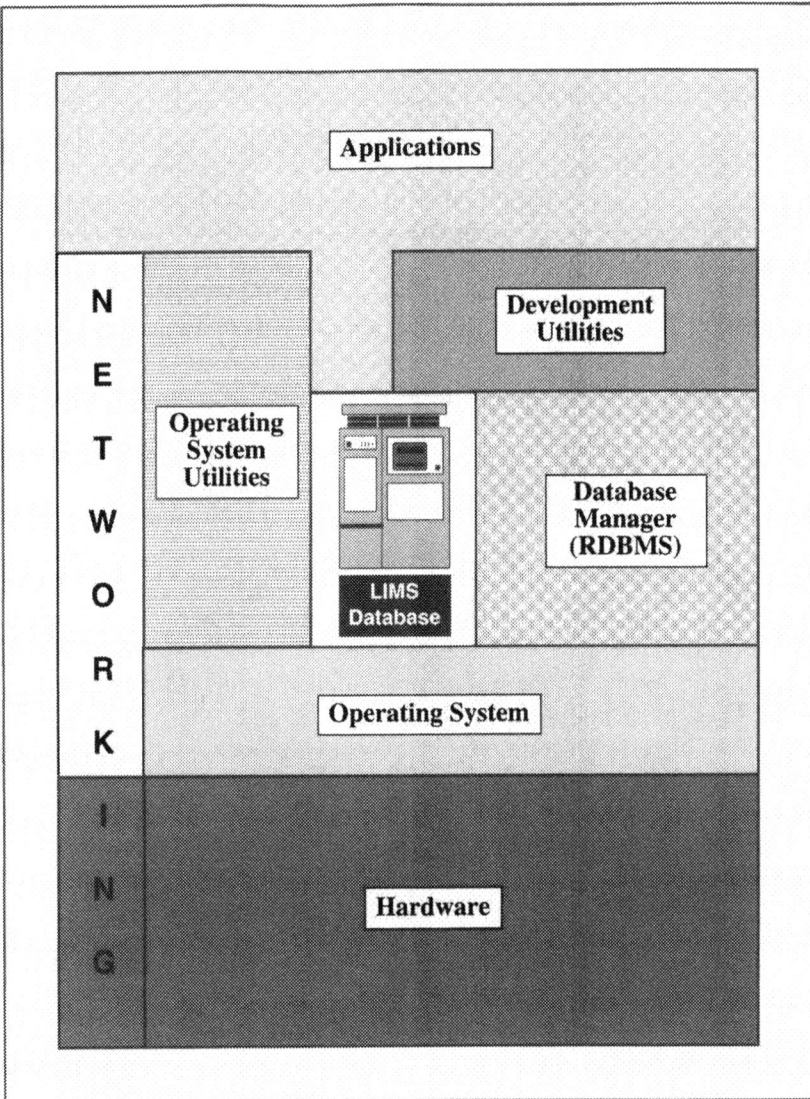

Figure 2-5 *Hardware and software architecture of a LIMS*

peripheral devices (such as terminals, printers, and disc drives), and elements such as cables and switches that connect the various parts.

The Computer Processor. The processor is the brains and control centre of the entire system. It acts on instructions included in the software to control the workings of all peripheral devices and other components of the system. By transmitting instructions in the form of prescribed electrical signals, it directs the functions of all elements of the systems hardware, software, and networking.

A computer processor may be composed of one or more components, each dedicated to a specific task. A low-end personal computer designed to serve one person at a time may contain only a Central Processing Unit and Disc Controller. Higher-end personal computers and workstations can fulfil much more demanding applications. They can, with the addition of networking, service several users at a time. Larger multi-user computers (such as servers, minicomputers, and mainframes) have many specialized semiconductor-based components. These are inherently designed to accommodate simultaneous support for multiple users and applications.

The diversity and sophistication of hardware within a computer processor increases with the size, complexity, and speed of the system. The processor appropriate for a particular use is based on the amount of data to be handled, the number of users, the complexity of the processing, and the number of peripheral devices to be handled.

Storage Devices. Storage devices serve as a repository of data and programs. They are holding areas where the computer processor retrieves what is needed to complete specific tasks. Storage devices are also used to store data created or modified by programs.

Disc drives are used for long-term storage of data and software which is generally accomplished by information transfer to and from magnetic media. The access time for the storage and retrieval of data ranges from milliseconds to seconds. Disc drives are based largely on moving parts which operate at high speeds. They physically deteriorate with time. This degradation is significantly accelerated when they must work in environments of unregulated temperature, dust, and corrosive vapours. Because of their mechanical nature, disc drives are the component of a computer system most prone to failure. The consequences are serious, since data can be lost.

Optical drives are noted for their large storage capacities. Data is encoded and stored on discs by a laser interacting with light sensitive media. The mechanism is similar to that of an audio Compact Disc. Like a disc drive, the storage media consists of a platter that rotates within the drive unit.

Optical drives are generally slower than disc drives in the retrieval of stored data. They do, however, have much higher storage densities and can hold extremely large amounts of data.

There are currently two main types of optical drive.

• *Read Only.* Information is permanently stamped on the optical disc. The data can only be retrieved. It is not possible to add or change what is stored on the drive. This technology is widely used for the distribution of software and data.

• *Write Once, Read Multiple.* Once written on the disc, information cannot be altered but can, thereafter, only be read. New information can be added until the disc is full. However, information cannot be deleted from these drives. This technology is useful for the permanent and unalterable storage of data.

Tape drives consist of magnetic media on a thin film. The tape is wound either on reels or in cassettes. Hundreds of feet of tape are contained in each unit. The tape is moved over a stationary sensor that sequentially reads or writes data.

Magnetic tape is generally a much slower means of storage than disc. Tape drives are normally used for the bulk transfer of data to and from disc drives; for backups, archiving data prior to deletion, and loading software.

Input Devices. Various devices are available to get data into the computer system. The way that you interact with these devices is controlled by software programs.

Terminals consist of a screen upon which characters or graphics are visually displayed. Software programs running on the computer create character or graphic images on the screen to which users respond, either by typing on a keyboard or by another device such as a mouse. After your input, the particular software program will react and follow up with additional characters on the screen, to which you must again respond.

Bar code readers provide an alternative means of entering character data through a terminal. They will read information printed in a coded format. The bar code format consists of alternating bars of varying thicknesses. Each letter or number is represented by a specific pattern of bars.

Bar codes provide an accurate way of reading sequential identifiers, such as the LIMS sample identifiers, into the computer. There are several types of bar code readers and bar code formats. Generally the technology works well if the bar coded information is printed clearly and is not soiled or marked prior to reading.

Scanners provide a way of entering images into the computer. A hard copy image is placed on a bed and the image is scanned by a light beam. Sensors translate the scanned image into a format that can be stored on the computer.

Printers. There are numerous types of printer, each with unique capabilities, quality of output, and costs. Some can only print characters while others can output both graphics and characters. Many create a print out by mechanically striking an inked ribbon such as dot-matrix and band printers. Laser printers deposit images on paper by electrostatic techniques, similar to a photo-copying machine. Departmental and work group printers are mechanically designed for constant use by many people. Other, less expensive, models are intended to be used only occasionally.

The paper-handling components of all printers are inherently mechanical in nature. Printers require regular maintenance and cleaning. During constant operation, they also produce a fine paper dust that can damage other computer components.

Communications Devices

Cables hold the physical components of the computer system together. They provide the pathway for movement of data and control signals to and from various devices of the system. There are various types and standards for computer cables. They differ by the manner in which data is transmitted through them, the speed at which information must move between devices, and the maximum distance between devices. The physical configuration of the cable connection and the communication standard between any two devices need to match in order to establish conductivity.

Networking hardware ties several computer systems together. They include specialized high-speed switches, protocol translators, and multiplexors. Networks require a sophisticated and tightly integrated web of specialized hardware, operating systems, network management software, and communications lines. A later section of this chapter discusses networks.

Interfacing hardware establishes data connectivity between laboratory instruments and the LIMS computer. Included are devices such as digitizers, signal filters, and instrument controllers. Interfacing hardware, in some cases, may be accompanied by specialized software for the acquisition and processing of data from instruments.

Software Components

Software is the instructions to be carried out by the physical hardware components. They are the means by which the computer system accomplishes its work. Users physically interact with hardware components such as terminals, bar code readers, and printers. However, the manner in which this interaction occurs is determined by detailed software instructions.

Software programs dictate how and which information is stored in the database, how calculations are performed, the format of reports, and how information is displayed on a terminal, and can even restrict selected individuals from carrying out specific functions.

Operating System. The operating system provides detailed instructions that control most basic functions of the computer system. It provides the fundamental coordination of all physical hardware components such as terminals, printers, disc drives, *etc*. The operating system also coordinates all the hardware and software components. The computer processor, disc drives, terminals, other physical devices, as well as the database, applications software, and custom programs all must be compatible with the operating system.

The Database. The database consists of laboratory data stored on the computer in a format and structure determined by the DBMS (Database Management System). The DBMS is a series of sophisticated programs and utilities that controls how data is stored and retrieved. It is the electronic equivalent of file cabinets with pre-designated locations for various items of information. There are subtle differences between the database and the DBMS: the database is the actual information that is stored, while the DBMS includes the structures, programs, and utilities that manage the storage of this information.

However, a DBMS is much more than the electronic equivalent of filing cabinets because it keeps track of where information should be stored. With a DBMS you can store or retrieve information without knowing where it should go. All you need to know are the descriptive names of the data that you are concerned with. The DBMS itself keeps track of the data's exact storage location.

Many DBMS products have features to recover data in the event of power failure or a mechanical breakdown of a disc drive. A transaction log records all changes made to the database since the last backup of the database. The DBMS uses the transaction log to restore information up to the point of failure. In the absence of transaction logging, all changes made since the last backup are lost and all laboratory analyses completed during this time must be re-entered.

Applications. The LIMS Application provides general laboratory functions. This includes common laboratory tasks such as sample log in, sample tracking, test scheduling, test data entry, test approval, and reporting. The exact functions provided and approach taken can differ significantly from one LIMS application to another.

A wide variety of *other applications* software can be on the system. This includes laboratory software packages for statistics, structural data management, word processing, spreadsheets, textual data management, image management, and document control. New vendors and new applications packages are constantly emerging. A few applications can perform specialized functions directly on the LIMS database. Others operate on their own specialized databases.

Utilities. Development utilities software packages are available for the development of applications to improve the efficiency and quality of the time-consuming software development process. LIMS vendors frequently use some of these tools for development and maintenance of their products.

System utilities operating systems are normally accompanied by special utility programs to monitor and manage system operations. Other system utilities can also be obtained from either the hardware vendor or third party sources. System utilities include specialized programs for the tasks listed below.

- *Text editing.* A text editor provides a quick way for the system manager and developers to enter software instructions. It is normally provided with every operating system. They generally have good text manipulation capabilities, but they lack the specialized character and document formatting abilities of the most basic word-processing programs.
- *Performance monitoring.* Performance monitors provide a way of keeping track of how hard various system components are working. It measures how the processor and other peripherals of the system are utilized. A good performance monitor can determine which hardware component or program is causing a slow response time. They provide supporting evidence justifying when and if hardware upgrades are warranted.
- *File back-up and restore.* These programs provide a means of copying selected (or all) files to an off-line storage medium such as magnetic tape. They also allow for restoration of data from off-line media back onto the system.
- *User security.* Security programs allow the systems manager to control access to parts of the system. The systems manager assigns each user an account and a password which are needed to use the computer. Furthermore, the systems manager can restrict what each person can or cannot do on the system.

Networking. Computer networks connect separate computer systems to one another. A network benefits the organization by allowing it to share data and hardware resources. A network consists of a series of highways through which data can move from one computer system to another. Specialized networking hardware and software coordinate the movement of data between the numerous hardware components. Networking software defines contact points for each hardware device and coordinates the transfer of data between the various points.

A network adds a layer of complexity, cost, and overhead to each computer connected to it. The complexity of establishing and maintaining a network increases with the total number of devices connected, the geographic dispersion of components, and the number of different hardware platforms included. A Local Area Network (LAN) connects devices in close proximity, within an office, a floor, or a building. A Wide Area Network (WAN) provides connections over metropolitan, national, or international distances.

There are a large number of possible network configurations. The concept of networking is sometimes intimidating and mysterious to the uninitiated. However, there is little magic involved for you to understand their use. It does, however, involve specialized technical expertise to install networks. Hardware vendors can provide detailed information on this subject. Other sources of information include networking specialists or network integrators.

Applications Architecture

The architecture of a system sets forth the applications required to support the business functions of the organization. It also considers the myriad of hardware, software, database, and networking environments in which each application runs. The applications architecture presents a distribution of hardware, data, and functions to meet the organization's needs. It specifies how the applications communicate with one another and how each component exchanges data with other systems within the architecture.

The functional model of a LIMS (Figure 2-3) is depicted as a series of functions interacting with a centralized LIMS database. The LIMS database may, in fact, be actually implemented through several physically separate databases and separate LIMS applications running on separate computers. The various functions surrounding the LIMS database may also be implemented on the same, or separate computer systems.

Other programs, besides the LIMS, are incorporated into the overall architecture, each targeted to a specific need. One possible applications architecture is shown in Figure 2-6. It presents a distributed approach and relies on many programs, each optimized and designed to address highly targeted and specific needs. It does not attempt to address all laboratory and enterprise needs through a centralized design in which all functions are provided by a single system. Past attempts at a centralized approach have not generally been effective.[5]

[5] A. Koller and G. Liesegang, *Am. Lab.*, 1992, **24:14**, 33.

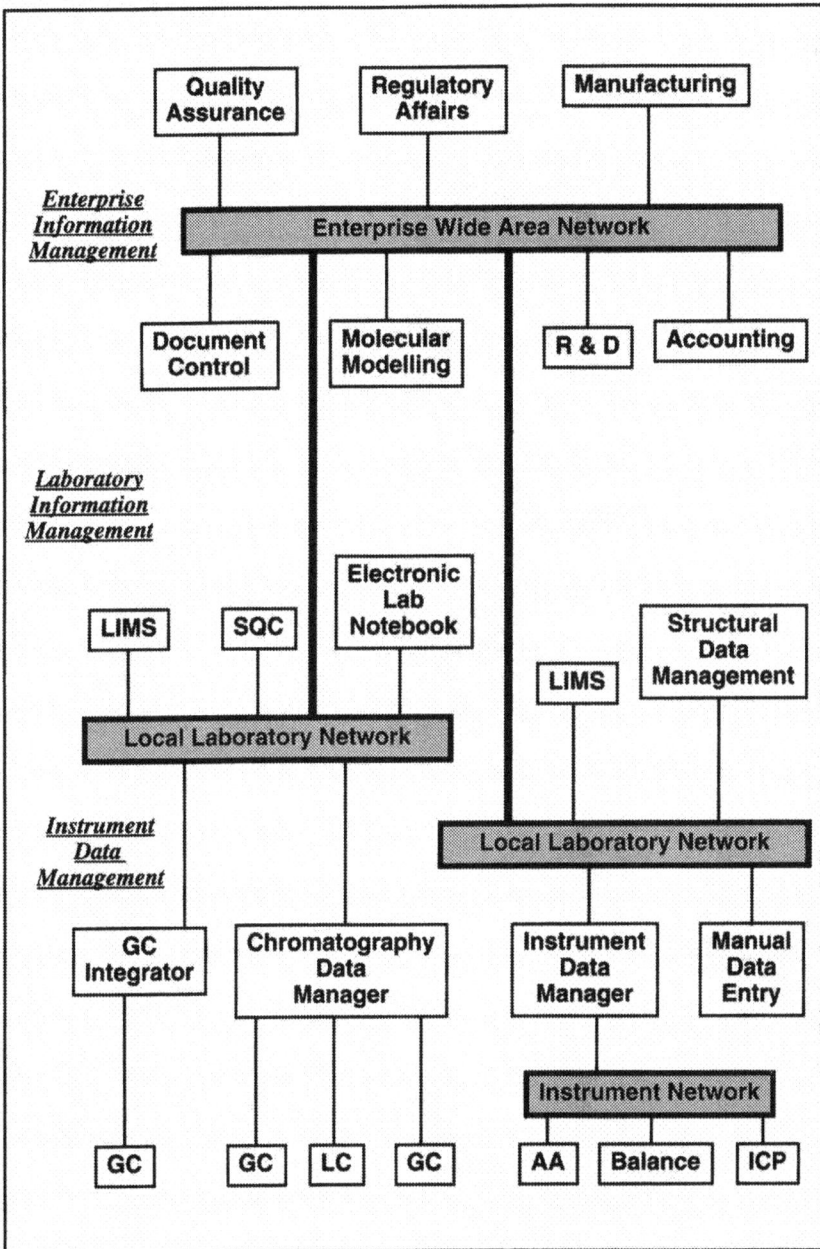

Figure 2-6 *LIMS applications architecture*

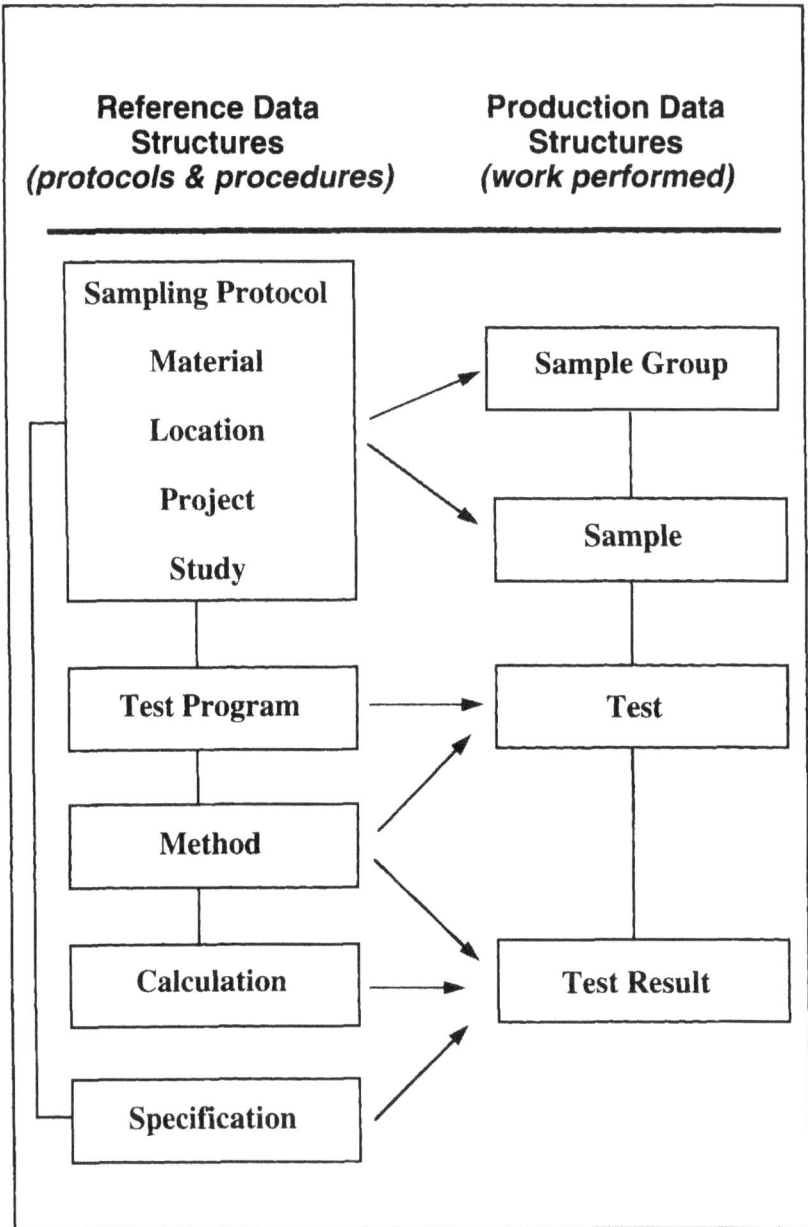

Figure 2-7 *LIMS data architecture*

Data Architecture

The data architecture sets forth the overall organization of the major LIMS data structures. In cases where the data is distributed over several separate databases, it also sets forth the physical data distribution scheme, and the mechanisms for data communications and synchronization.

A LIMS database is normally organized into two major segments, reference data structures and production data structures. One example is provided in Figure 2-7.

The reference data structure describes laboratory protocols and procedures. It includes sampling schedules, testing schedules, test programs, projects, test methods, calculations, and specifications. Reference structures are also called 'template' or 'static' data. They represent what the laboratory should do and the information that should be captured as work proceeds.

Work actually performed by the laboratory is represented by production data structures. These are also called 'instance' or 'dynamic' data. They represent information created and updated as an outcome of sample processing and analytical testing. This includes data on samples, tests, and results.

Before the LIMS can be used, the laboratory needs to populate the reference data structures with values corresponding to their own operating procedures and protocols. Henceforth, the LIMS programs enforce the laboratory's procedures in its analytical activities.

CHAPTER 3

Techniques for Understanding Laboratory Operations

Each laboratory performs many sophisticated and interrelated tasks to fulfil its contribution to the overall enterprise. This includes activities that directly involve sample handling and analytical data creation. It also includes additional work to support various testing and administrative activities. Often, the laboratory analyses test results to assist others with decisions or action plans based on those results. Laboratories are also often involved in the design and execution of experimental protocols, manufacturing processes, sampling schedules, monitoring programs, test methods, and specifications. The overall complexity of laboratory operations is generally underestimated. In many instances, the quantity of data produced and managed by the laboratory equals or exceeds the total data generated by all other groups in the organization!

Various methodologies are available to assist in understanding operations. Many automation projects fail because the needs and constraints of the targeted organization are not well understood prior to implementation. From the perspective of technical hardware and software, the installed systems may be technically elegant. However, the most advanced state of the art technology is of little use if it fails to meet the laboratory's fundamental business objectives and if it does not blend with its day to day operating needs.

1 Characteristics of Laboratory Operations

There are three dimensions to the operation of any system that need to be fully understood prior to automation: its process, data, and temporal (time-dependent) behaviour. Aspects of each organization's operations can be visualized as an irregularly shaped body as illustrated in Figure 3-1. Processes include the various actions and tasks that are performed by the organization. The data involves the exact information used or generated by all processes. Temporal behaviour includes the numerous time-critical relationships among the various processes and data within the organization. Given the laboratory, or laboratories, planned for automation, where do you start in understanding its operations?

Unfortunately, most techniques typically focus on one of these dimensions, because they were intended for special types of systems. The systems that are

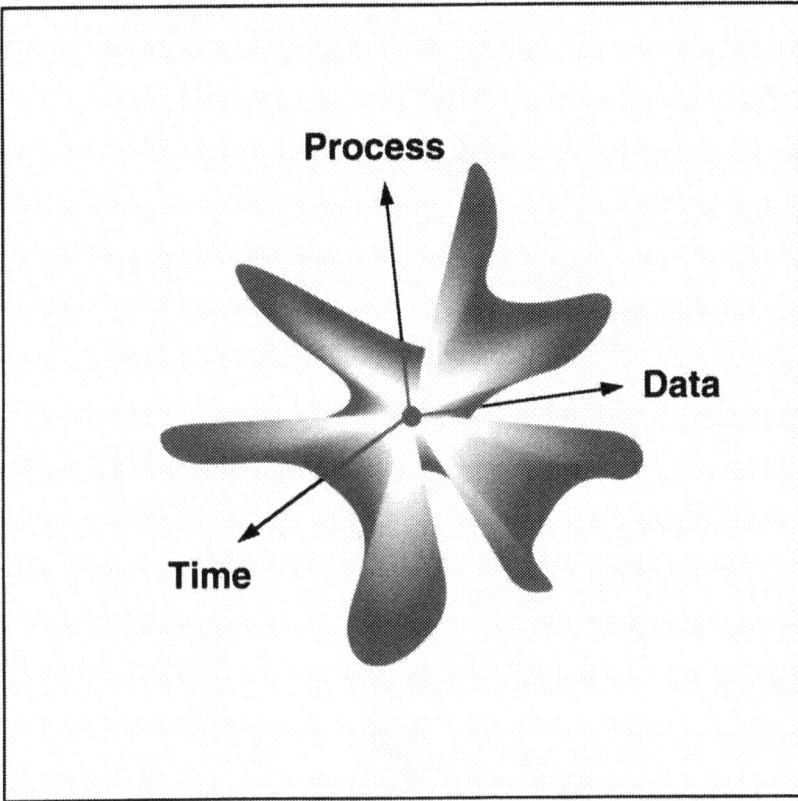

Figure 3-1 *Dimensions of each organization's behaviour*

predominantly complex in only one dimension are systems well served by these methodologies.

- Most business applications have high levels of functional complexity.
- Real-time control and data acquisition systems have highly precise time-critical requirements.
- Flexible enquiry and reporting systems have high levels of data complexity.

Our organizations and the systems serving them are getting increasingly complex in more than one dimension.

Laboratories have always required significant attention in all three dimensions: process, data, and temporal characteristics.

- Complex processes are involved in the acceptance, scheduling, testing, review, validation, release, and reporting of samples.
- The actual data managed by laboratories often involves information needed to work on samples, but also descriptive information provided by others for reporting and testing.

- Many tests have time-critical components. Deviations from prescribed time windows means that tests and their associated results are invalid. Laboratories are also filled with real-time data acquisition and data analysis systems with extremely rapid temporal requirements.

2 Techniques for Understanding Operations

Techniques are presented for representing the following aspects of your laboratory:

- context diagrams for determining the scope of the laboratory's interactions with external groups;
- functional decomposition diagrams for a hierarchical summarization of all laboratory processes;
- flow diagrams to represent detailed task and process relationships;
- event diagrams for depiction of time-dependent behaviours and relationships; and
- data models to detail the various data elements used or created by the laboratory.

Each of these techniques is based on established systems modelling method-ologies. Each model is nothing more than a set of conventions for representing the organization's behaviour and its needs. Graphics depict relationships and behaviour. Diagrams that are clear and easy to understand facilitate the visualization and precise definition of operations. Graphic elements are complemented by text which details the individual elements on the graphical models. The conventions and their corresponding graphical models are easier to create, easier to understand, and are more concise than descriptions solely based on long textual prose. They facilitate communications between those involved with automation and those intimately familiar with the inner workings of the organization. Without a common understanding, the system may be designed to address perceived operations, rather than the actual functions of the laboratory.

A detailed discussion of modelling methodologies is discussed in Chapter 7. The purpose for their inclusion in this section is to establish conventions to facilitate a discussion of the varied characteristics of laboratory operations to be covered in the next section.

Establishing Boundaries: The Context Diagram

A context diagram establishes the role of the laboratory within the overall enterprise. It sets forth work that flows into the laboratory as well as work that flows out. It establishes the various organizational units that affect, or are affected by, the laboratory.

The purpose of the context diagram is to set forth the domain or scope of laboratory services. The organization under analysis, in this case the laboratory, is graphically shown as a rectangle in the middle of the diagram (Figure 3-2). Groups outside the laboratory, referred to as terminators or external agents, surround the laboratory. These external agents may be a person, a group of people, a government agency, or a department within or external to the

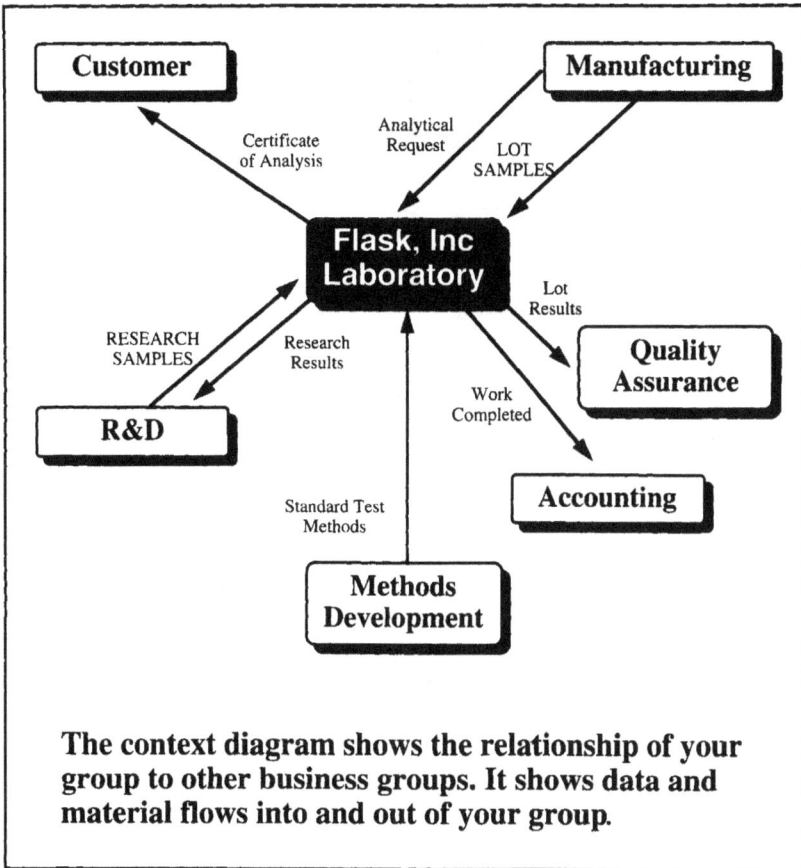

Figure 3-2 *Example of a context diagram*

enterprise. External agents are outside the direct control of the laboratory, but they do represent entities with which the laboratory must interact.

Flows to and from the laboratory are shown as connecting lines. Arrows at the ends of the lines indicate the direction of flow. Each flow consist of movement of data or materials such as samples, paperwork, verbal notification, or a telephone call. Flows may also originate from another computer system.

A typical laboratory may be much more complicated than this example. There may be, in fact, scores of external agents and extremely complex flows. The principal objective of the context diagram is to establish the major players and the flows into and out of the laboratory. Each person that you interview may have a different perspective on the laboratory's context with the outside world. Their view is moulded by their own job responsibilities and their perspectives are shaped by their personal interactions with others. The essential context diagram emerges through a synthesis and consolidation of information gathered from several sources.

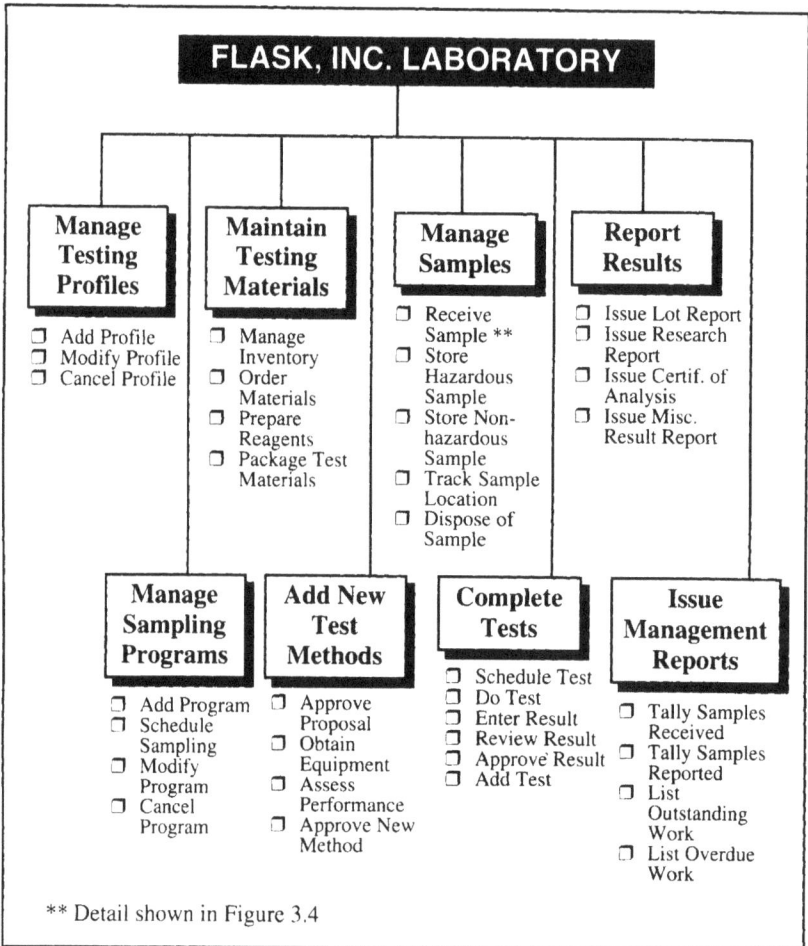

FLASK, INC. LABORATORY

Manage Testing Profiles
- Add Profile
- Modify Profile
- Cancel Profile

Maintain Testing Materials
- Manage Inventory
- Order Materials
- Prepare Reagents
- Package Test Materials

Manage Samples
- Receive Sample **
- Store Hazardous Sample
- Store Non-hazardous Sample
- Track Sample Location
- Dispose of Sample

Report Results
- Issue Lot Report
- Issue Research Report
- Issue Certif. of Analysis
- Issue Misc. Result Report

Manage Sampling Programs
- Add Program
- Schedule Sampling
- Modify Program
- Cancel Program

Add New Test Methods
- Approve Proposal
- Obtain Equipment
- Assess Performance
- Approve New Method

Complete Tests
- Schedule Test
- Do Test
- Enter Result
- Review Result
- Approve Result
- Add Test

Issue Management Reports
- Tally Samples Received
- Tally Samples Reported
- List Outstanding Work
- List Overdue Work

** Detail shown in Figure 3.4

Figure 3-3 *Example of a functional decomposition diagram*

Once the context of the laboratory has been established, the diagrams are useful for establishing the scope of the planned LIMS.

Obtaining an Overview of Processes: The Functional Decomposition Diagram

A functional decomposition diagram graphically depicts a comprehensive and hierarchical inventory of all laboratory business functions. An example is shown in Figure 3-3. Unlike a context diagram which concentrates on external interactions, the functional decomposition concentrates solely on work within the laboratory.

Each business function is described by a verb–object phrase. The detailed processes and tasks underlying each business function are amplified by flow diagrams.

It is important to note that the functional decomposition represents a high-level functional breakdown of work within the laboratory. It is not a reflection of the laboratory's organizational chart. It serves to reflect the essential business functions undertaken. How these functions are accomplished and who is responsible for them changes with each reorganization. A properly executed functional decomposition is independent of the organizational structure, personnel, or technology used to accomplish work. Reorganization of your laboratory should not affect its functional decomposition. Changes to the functional decomposition occurs only when work responsibilities are transferred into or out of the organization.

Understanding Process Relationships: Flow Diagrams

Details of each process shown in a functional decomposition can be amplified through flow diagrams. For example, elaboration of the process called Receive Samples from the functional decomposition (Figure 3-3) is detailed in the data flow diagram of Figure 3-4.

Flow diagrams can be used to illustrate relationships between processes from the functional decomposition as shown in Figure 3-5.

Flow diagrams are useful for understanding the movement of information and samples through the laboratory. Names commonly used are work flow, data flow, and sample flow diagrams. The next section presents conventions and components of flow diagrams.

Flow Diagram Conventions

The primary focus of a flow diagram is the various processes and their relationships to one another. Conventions applicable to flow diagrams are shown in Figure 3-6.

Processes

Processes are graphically shown as circles. A process name is contained within each circle. Each process name consists of a verb–object phrase describing the action performed.

The principal objective is to understand what the laboratory does, independent of how the process is currently completed. The names of groups, departments, organizations, or individuals do not serve as appropriate process names.

Stores

Stores are shown as two parallel lines. They are packets of information or materials at rest. Stores may consist of a file cabinet, a shelf, a stack of paper on someones desk, or a computer database.

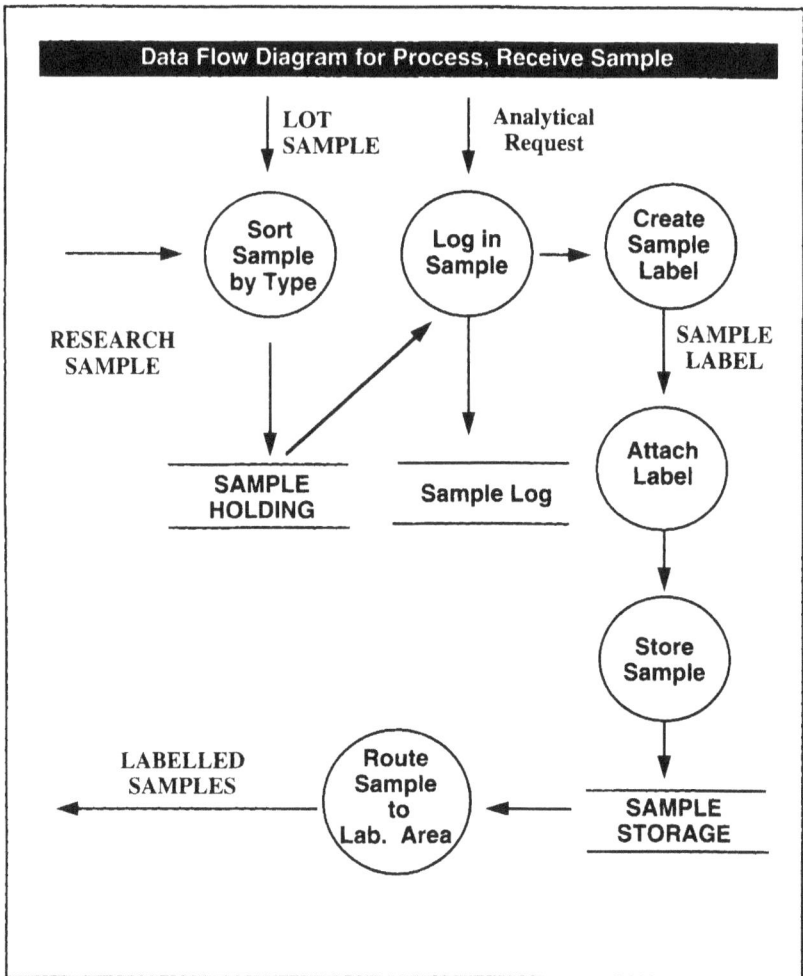

Figure 3-4 *Example of a data flow diagram*

Flows

Flows are shown as arrowed lines. They describe the movement of packets from one part to another. The direction of the arrow indicates the direction of the flow. Flows consisting of data or materials are accompanied by the name of the data or materials. For those lacking a descriptive name it implies a sequence or chronology from one process to another.

Data Flows. Data flows are labelled with a name in mixed-case letters. The first letter is in uppercase and the remaining letters of the name are in lowercase. Examples of data flows are Worksheet, Sample Label, Certificate of Analysis, and Turnaround Report. Data flows consist of packets of information. They are physically packaged in a variety of media such as paper, verbally, or electronically.

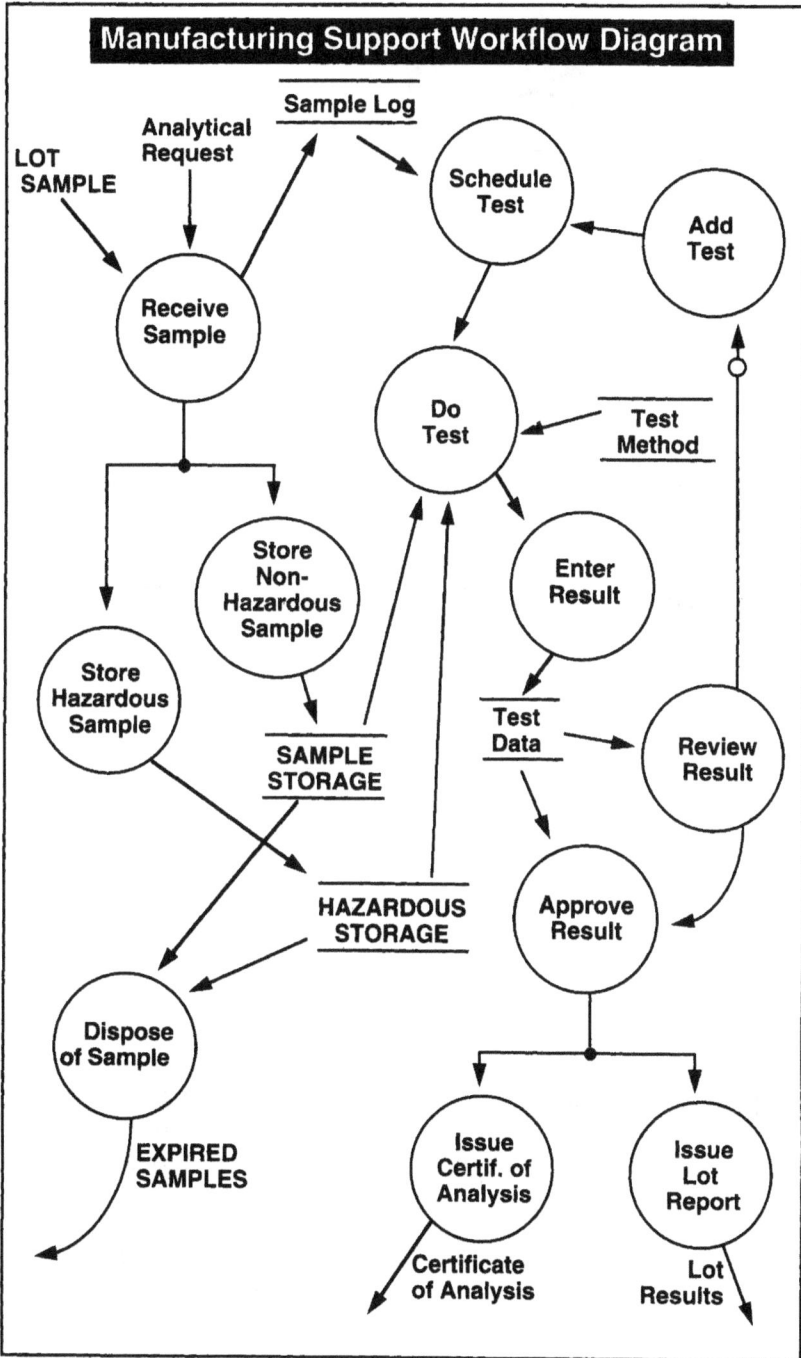

Manufacturing Support Workflow Diagram

Sample Log

Analytical Request

LOT SAMPLE

Schedule Test

Add Test

Receive Sample

Do Test

Test Method

Store Non-Hazardous Sample

Enter Result

Store Hazardous Sample

SAMPLE STORAGE

Test Data

Review Result

HAZARDOUS STORAGE

Approve Result

Dispose of Sample

EXPIRED SAMPLES

Issue Certif. of Analysis

Issue Lot Report

Certificate of Analysis

Lot Results

Figure 3-5 *Example of a flow diagram depicting process relationships from functional decomposition diagram (Figure 3–3)*

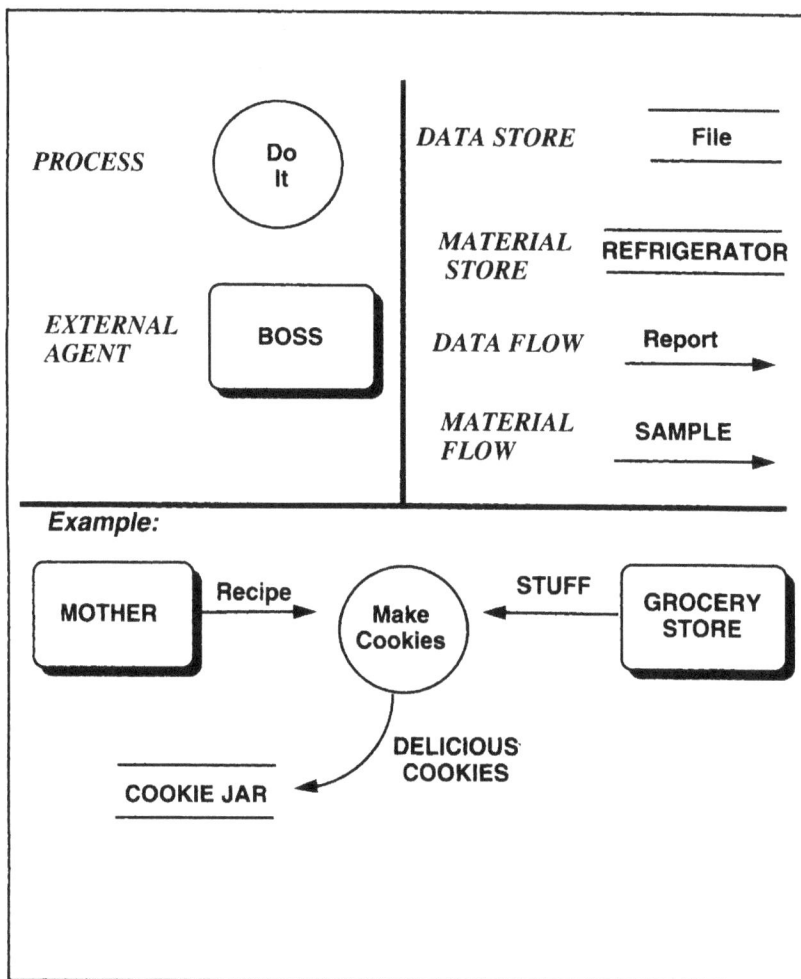

Figure 3-6 *Symbols used in flow diagrams*

Material Flows. Material flows are labelled entirely with uppercase letters. Material flows consist of tangible, physical objects processed by the laboratory. Examples are RAW MATERIAL SAMPLE, SAMPLE ALIQUOT, CALIBRATION STANDARD, and REAGENT BLANK.

Sequence Flows. The connecting lines for sequence flows are not labelled. They illustrate chronological relationships between processes. All data and materials may be passed through processes linked by sequence flows.

Optional Flows. Optional flows indicate that a process may or may not occur following its predecessor. This relationship is shown by a circle next to the arrow on the line connecting the two processes. The conditions under which the flow

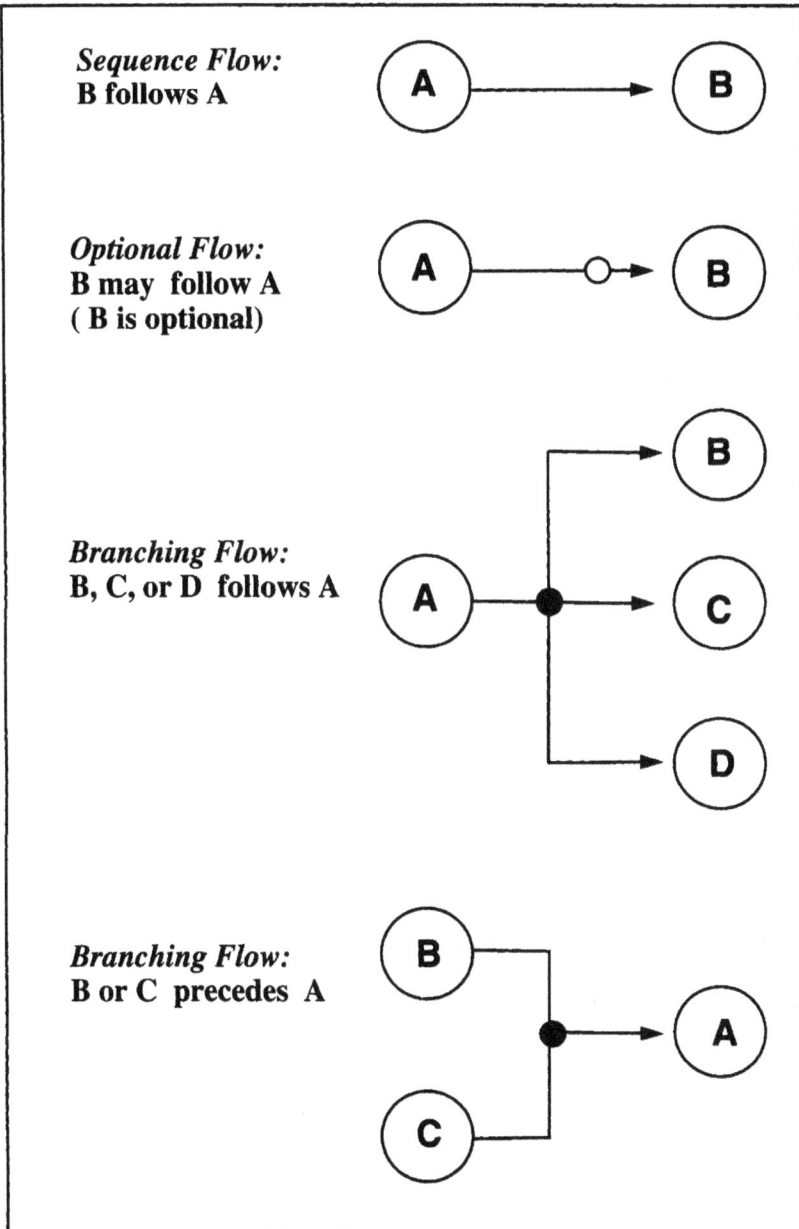

Figure 3-7 *Diagramming flows*

occurs, or does not occur, is detailed in the textual process descriptions. Alternatively, the condition may be written on the flow line. In these cases, the conditional statement should be written in italics to ensure that it is not confused with data or material flows.

Branching Flows. A process may be related to a group of subsequent or preceding actions. This is indicated by a branching line with a filled-in circle at the node of the branch (Figure 3-7).

Expanding Detail

Many flow diagrams in practice have a large number of processes and would require vast charts of paper. To ease presentation, flow diagrams can be divided into components that fit on normal-sized paper. A complex process may be expanded onto another more detailed flow chart on a separate page. Processes for which a more detailed flow diagram is available is marked by three dots above the process name.

Fundamental processes, also known as procedures, are the lowest level shown in flow diagrams. Each procedure is detailed by a brief textual description of no more than one or two pages which includes the following.

- *Name.* This should be the same name as the flow diagram.
- *Purpose.* The objective and reason why the process is performed.
- *Description.* A terse description of what occurs during the activity. It refers to any computer programs, forms, or reports involved in the activity.
- *Responsibility.* Who or which job role is responsible for the work?
- *Mode.* Is the process automated or manual?

Certain individuals, especially those with a background in programming, may be tempted to pursue too much detail in the creation of flow diagrams. Remember that, at this point, the principle objective of creating flow diagrams is to assist in the understanding of laboratory operations, not to design software.

Alternative Flow Diagramming Conventions

Several flow diagramming conventions are available.[1-3] Alternative methods for representing processes are ovals or rectangles with rounded corners. Data stores may also be shown as rectangles. The actual conventions used are not as important as consistency in the shapes used. The conventions that you adopt should also be clearly explained to prevent confusion.

Understanding Temporal Relationships: Event Diagrams

Event diagrams are used to understand the time-dependent behaviours of an organization. They complement flow diagrams and other models to improve the understanding of laboratory operations. Events trigger the activation of discrete processes. For example, the event 'telephone rings' activates the process 'answer phone'. Processes generally remain in an inactive or idle state without a

[1] E. Yourdon, 'Modern Structured Analysis', Yourdon Press, Englewood Cliffs, NJ, 1989.
[2] C. Gane and T. Sarson, 'Structured Systems Analysis: Tools and Techniques', IST, New York, NY, 1977.
[3] J. Martin, 'Recommended Diagramming Standards for Analysts and Programmers', Prentice-Hall, Englewood Cliffs, NJ, 1987.

triggering event. An action is not generally initiated, such as answering the phone, without a reason to do so.

Events are shown as large arrows on diagrams. There are three basic types of events: temporal, conditional, and flow.

- Temporal events occur at regular time intervals. Examples include time to start work, time for lunch, at the end of each week, at the beginning of each month, at 1:00 pm each Friday. The exact time may be defined via policy or it may be left to the discretion of individuals.
- Conditional events activate processes based on the occurrence of a specified condition. Examples include a feeling of hunger, sample storage area full, a need to use the bathroom. The exact definition of the condition may or may not be formally defined through a policy.
- Flows of material or data can also activate events. Examples include the arrival of samples, the arrival of a bill, the arrival of a test request. The timing for arrival of the flow is determined either through a preceding process or by an organization outside the laboratory.

Often the organization itself does not fully comprehend how it responds to events. In one case, almost everyone in the laboratory perceived that the arrival of a sample triggered the process for logging samples into the laboratory. In actuality, the people involved were responsible for many other tasks. Samples were only logged once each day, at 10:00 am. Work submitted to the laboratory after the scheduled time was not acknowledged until the following working day. Clients of the laboratory were perplexed by the management's claim of a 1 day turnaround, when they knew that it normally took 2–3 days to obtain results. In this example, the process was triggered by a temporal event (at 10:00 am of each work day), rather than a flow (arrival of sample).

One type of event diagram is illustrated in Figure 3-8. This is an elapsed time analysis. It studies the contribution of various processes to the overall time needed for the laboratory to respond to a service provided to an external group. The diagram shows the relationships between the various events, processes, and elapsed time intervals for work in a hypothetical laboratory. Included are several processes carried out by groups outside the laboratory (such as 'Acquire Sample', 'Send to Lab', and 'Apply Result'). The diagram shows idle queue times between processes during which no progress occurs. This type of diagram is useful for identifying bottlenecks. It also illustrates the various components that contribute to overall laboratory service turnaround time. Elapsed-time analysis diagrams serves as a basis for exploring automated or manual performance improvements to the current system.

A state transition diagram illustrates the life-cycle transition of key laboratory entities such as sample, test, sampling program, specification, or method. Figure 3-9 is a hypothetical state transition diagram for a sample as it proceeds through the laboratory. The defined status states are indicated by the horizontal lines. Processes that move the sample from one state to another are shown in circles. In this example, the process called 'Archive Data' moves the sample from states of 'cancelled' or 'reported' to 'archived'.

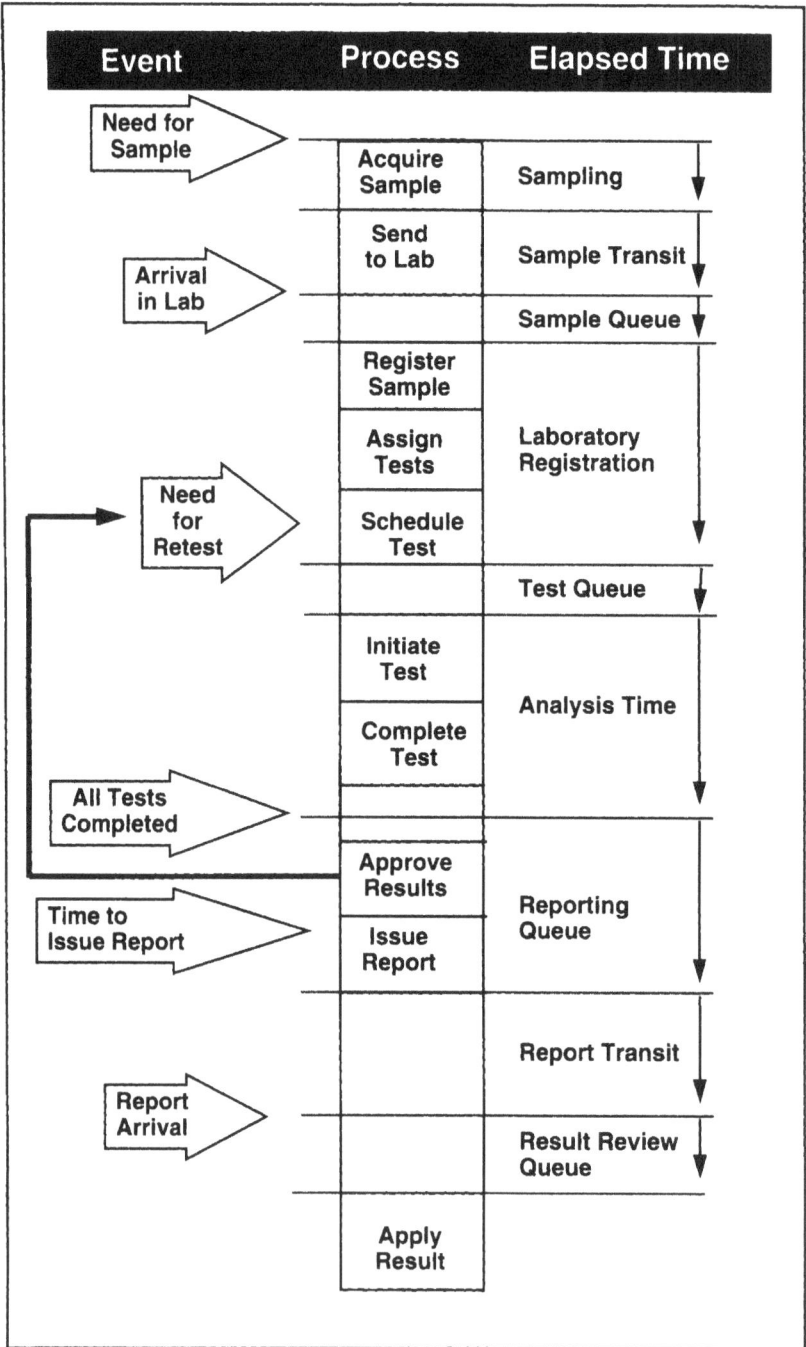

Figure 3-8 *Elapsed-time diagram*

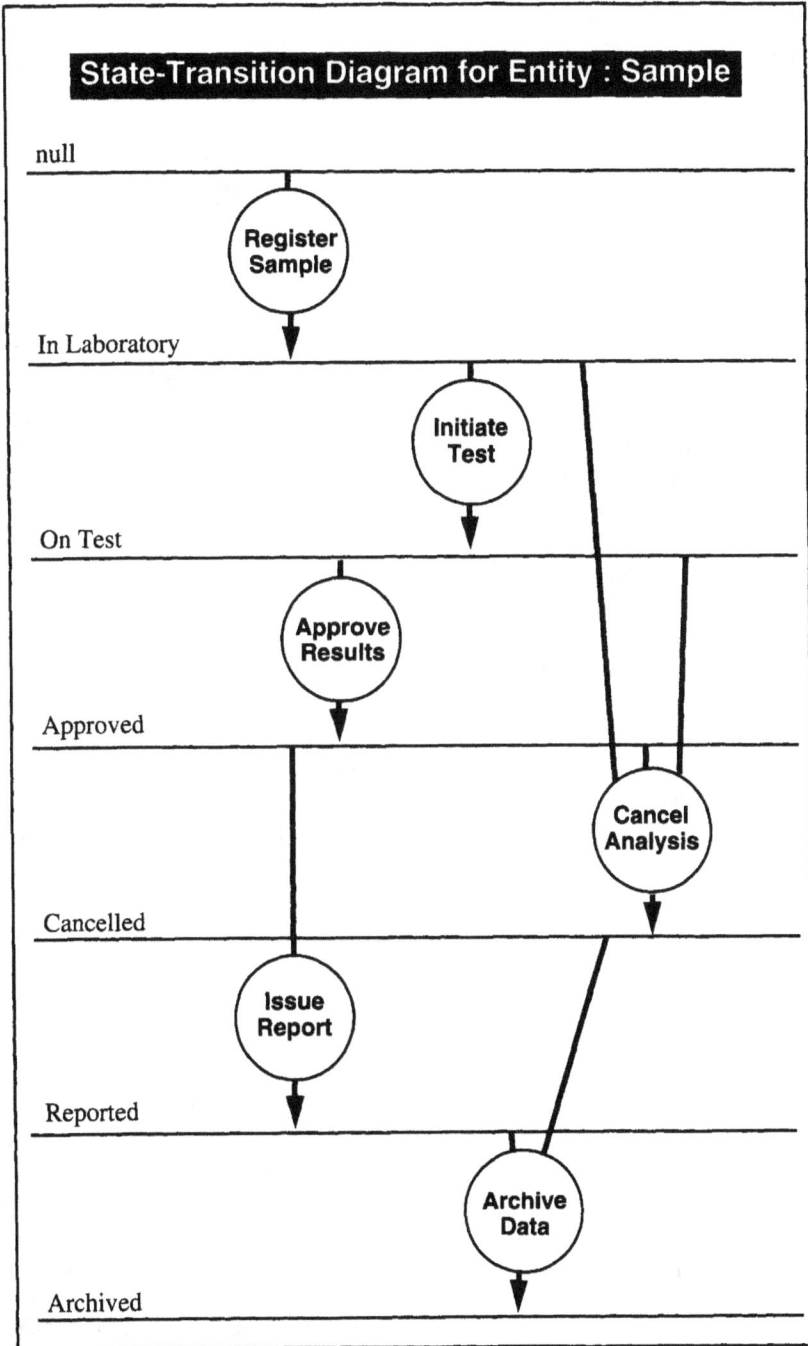

Figure 3-9 *Example of a state transition diagram*

Understanding Data Elements: The Data Dictionary

The data dictionary is an organized listing of all pertinent data elements, along with precise definitions understandable to those within and external to the laboratory. The data dictionary has nothing to do with the way that information will eventually be stored on the computer; it is simply an inventory of the data and its current meanings. However, the data dictionary does serve as a basis for evaluating appropriate database designs.

Organizations frequently lack a common base of definitions for data used in their daily activities. Laboratories frequently operate under a 'common understanding' of what things mean. Exact definitions are never established. Unfortunately, various people and groups within the laboratory carry on with slight to substantially different interpretations. The result can be data which is inconsistently generated and inappropriately interpreted. The process of creating a data dictionary generally identifies issues surrounding terms that are currently vaguely defined or inconsistently applied. It provides an opportunity for the laboratory to recognize and, if warranted, address any internal deficiencies prior to automation.

The basic information needed to create a data dictionary is found in existing reports, forms, notebooks, and other printed or written records of the laboratory. The various documents should be gathered and catalogued. Definitions are obtained for each element of data on every form along with a brief description of how it is used.

Identification of the various data elements and their definitions needs to involve the individuals who actually use the forms each day. Their involvement is important for several reasons.

- Often it is found that critical information is written in the margins or on the back page. The preprinted captions on the forms do not provide a comprehensive inventory of its data content.
- Handwritten marks, initials, and dates often have meanings that are not obvious.
- Many laboratories do not update their forms and reports as their procedures change. The spaces allocated for data are no longer used or they may be used for a different purpose than originally intended.
- A data element may have several meanings, depending on the situation.

Data elements from all sources studied are then consolidated and unique names are assigned. An element entitled 'sample weight' on a sample submission form probably has a different meaning to one with the same title on an analysis worksheet. For sample submission, it may mean the total weight of the sample upon submission to the laboratory, whereas on an analysis worksheet the term probably means the weight of the material tested either before or after it has been subject to some form of physical or chemical manipulation. In this event there are two data elements that may be named 'sample weight upon receipt' and 'sample weight after incubation'.

Understanding Data Relationships: The Data Model

A data model maps out all the types of laboratory information and their relationships to one another. Some, or all, of the data routinely processed by the laboratory may be automated through the LIMS. Other information can be handled through other systems or they can continue to be processed manually. The data model serves as a basis for determining aspects of laboratory data to be served by the LIMS database. It also serves to ensure that the computerized system is designed and configured in consistence with the needs of the laboratory.

Entities

The first step is to categorize information into logical groups called entities. An entity may be a physical object or an action performed. Entities are usually described as nouns and verbs. Example entities are sample, method, instrument, result, and specification.

Attributes

Attributes detail the unique characteristics of each instance of an entity. For example, the entity Laboratory Analyst can have instances called John, Paul, Peter, and Mary. Attributes of each include their job title, department, date hired, birthdate, telephone number, and home address. The data model is only concerned with attributes necessary to support the various business processes of the organization. Hence many attributes of each Laboratory Analyst, such as their height, weight, dietary preferences, and hobbies, are not included in the laboratory's data model.

Normalization

In some instances, several attributes are repeated when describing multiple occurrences of an entity. For example, a Laboratory Analyst may be described as the department to which they belong along with its code, manager, and location. All Laboratory Analysts within the same department have an identical code, manager, and location. Through a process called data normalization, the department is broken down into an entity unto itself. In doing so, attributes describing it are listed only once and are not repeated for each analyst. An example of this is illustrated in Figure 3-10.

Diagramming Relationships

The numerical relationships between entities is called their cardinality. An entity–relationship diagram depicts these relationships with the following graphical conventions:

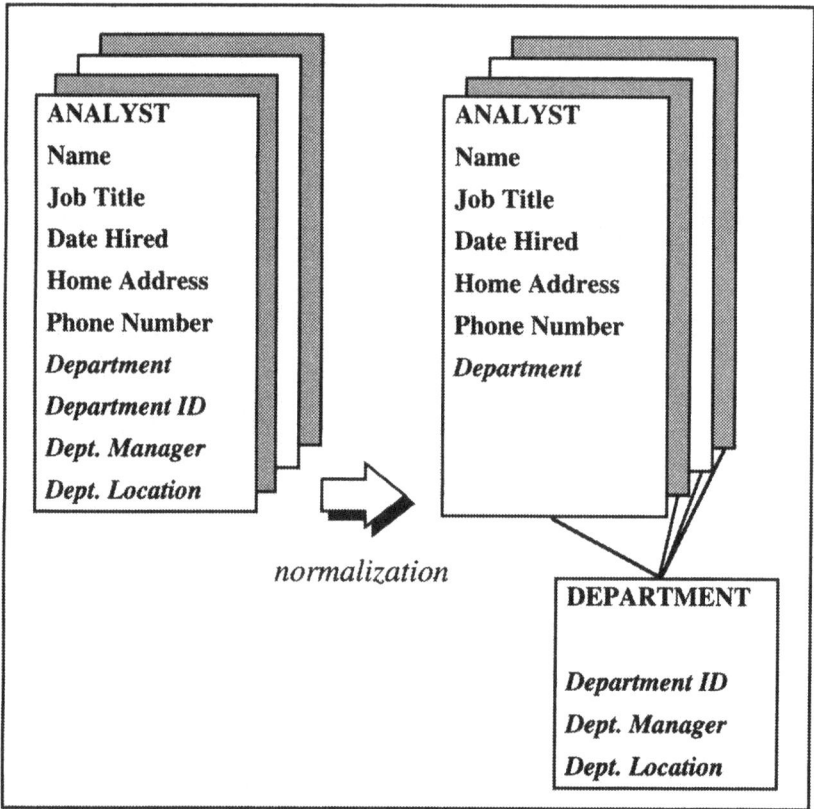

Figure 3-10 *Example of data normalization*

- an entity is a rectangular box;
- a relationship between two entities is shown as a line connecting the boxes; and
- the numerical relationship is indicated by symbols at the opposite end of the connecting line as illustrated in Figure 3-11.

It may be found that the laboratory data model has many entities that cannot easily be handled by a single diagram. Such a diagram would be too cluttered and difficult to understand. The data model can be simplified by breaking it up into several separate fragments, each representing a logical group of entities. A single entity may belong to several fragments simultaneously.

One possible grouping is to create separate diagrams for entities related to:

- testing methods and procedures;
- sampling protocols;
- sample testing and test results;
- test administration and scheduling; and
- laboratory administration.

An example entity–relationship diagram is shown in Figure 3-12.

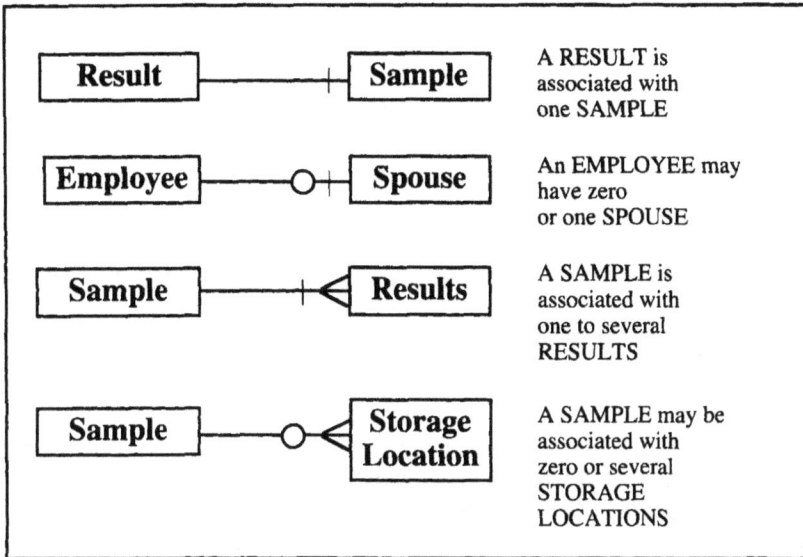

Figure 3-11 *Entity–relationship diagram cardinality symbols*

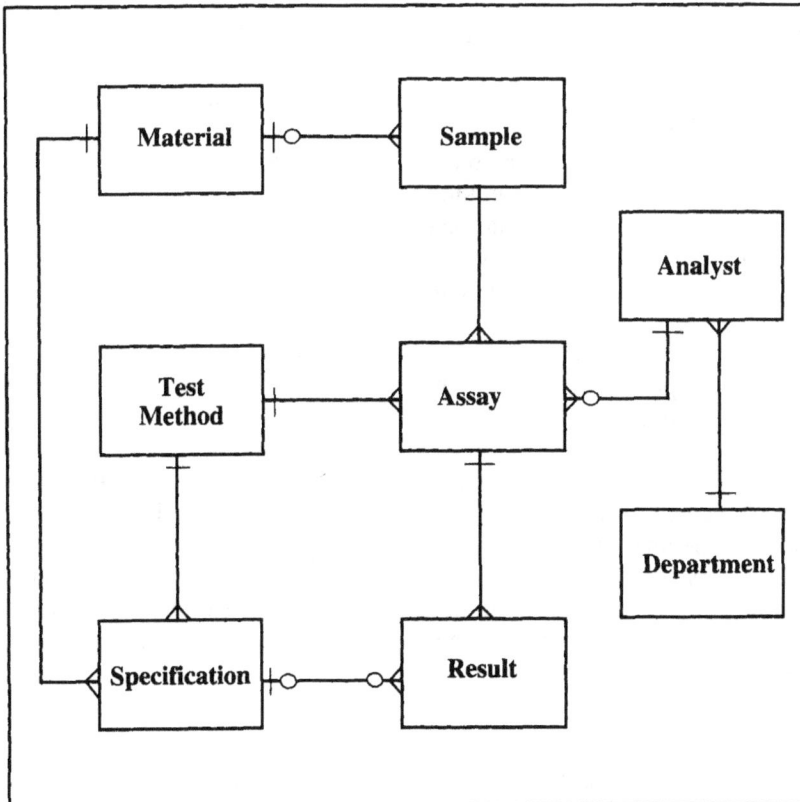

Figure 3-12 *Example of an entity–relationship diagram*

CHAPTER 4

Laboratory Roles and Internal Interactions

The importance of laboratory analysis to modern industrialized economies cannot be overstated. The current state of our industrial and medical capabilities has been catalysed by advances in our analytical capabilities. Data from our research laboratories result in newer and increasingly innovative products. These advances are imperative to our effective competition with developing nations who have low labour costs and an abundance of natural resources. Laboratories also support the manufacture and servicing of almost everything we use. They determine the adequacy and quality of our existing products or services for use by customers. Laboratory testing ensures the relatively high standards that we often take for granted in our health care, the environment in which we live, and the modern amenities afforded by today's technology. Laboratory testing services are critical to the operations of various enterprises within our modern societies. Without them, our economy and our personal welfare would rapidly deteriorate. In the United States alone, 250 million measurements are performed each day, at a cost of 3.5% of the Gross National Product.[1]

This chapter presents the important roles of laboratories in an organization. The first section presents examples of several different laboratories, their role in the overall organization, and the processes needed to support their day to day activities. These examples illustrate a few, but by no means all, settings. The second section presents the various dimensions, manifestations, and uses of data within a laboratory.

A common misconception is that all laboratories have similar information management needs. It is true that all laboratories are organizations dedicated to the measurement of specific analytical properties of samples. However, what is measured, the analytical technology utilized, and why the measurements are needed differs significantly from one organization to another. These differences affect the processes and information that must be managed. A LIMS implementation approach that is appropriate for one laboratory environment may be totally inappropriate for another.

---

[1] J. G. Grasselli, 'Analytical Chemists: Problem Solvers in the Year 2000', Society of Analytical Chemists of Pittsburg, Pittsburgh, PA, 1991.

1 Laboratory Roles and Processes — Examples

This section describes the operations of three different laboratories: research, quality control, and analytical services.[2] It presents generic and idealized representations of each type of laboratory. In the real world, laboratories are much more complex. The purpose of this section is to present a sampling of different environments to illustrate how laboratories differ. It serves as a starting point for understanding the characteristics specific to your organization. The examples utilize context and flow diagrams previously discussed in Chapter 3.

Research and Development

Research laboratories increase an organization's knowledge base for new technology. Their innovations lead to new or improved products, processes, or services. Research laboratories transform the fundamental concepts of science into practical applications and new products. They also routinize new technologies so that they can move from the laboratory to mass production. Progress on projects is not gauged by the number of samples analysed, but on the utility of the new knowledge generated as an outcome of work performed by the laboratory.

Figure 4-1 shows the context diagram for one type of work typically conducted by research. Different diagrams would be needed to represent research support for marketing, product development, engineering, and manufacturing.

In this example, projects are funded through several sources such as corporate management, private foundations, or government agencies. Funding is provided to improve the technical virtuosity of the enterprise. Periodically, reports are issued to summarize findings and developments. Evaluation of these reports determine if the technology is promising and if other studies are warranted. They also determine if changes in the project's overall approach and direction are called for.

An overview of a research laboratory is shown in Figure 4-2. For each project assigned, the feasibility is determined based on the organization's knowledge and past experiences. At this point the constraints are identified and appropriate resources are allocated.

A study is designed to verify assumptions and to test hypotheses. The study stipulates analytical procedures, sampling requirements, and other conditions. Protocols for the study may come from standard methods, but in most cases the method of analysis is developed as a part of the project.

Samples are created and numerous observations and conditions are recorded. The samples are then tested and analytical data is generated. This is then processed, calculated, or otherwise manipulated through a variety of statistical, graphical, or other analysis methods (automated or manual). Conclusions are then formulated, summarized, and reported.

Research is inherently a risky and expensive proposition. It involves a high

---

[2] A. S. Nakagawa, 'Contrasting Logical Information Flow Models of Analytical Services, Quality Assurance, and Research Laboratories', Pittsburgh Conference on Analytical Chemistry and Applied Spectroscopy, New York, 1990.

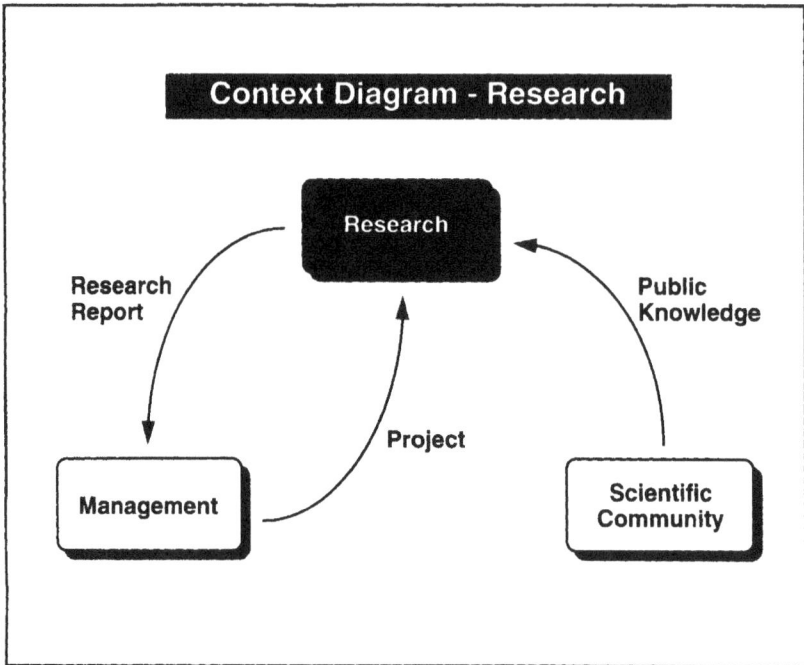

Figure 4-1 *Example of a context diagram of a research organization*

degree of uncertainty, sophisticated apparatus, and highly educated, creative, and expensive people. Much of research involves highly unstructured and undefined processes which defy automation. Innovation generally relies on the creative and intellectual abilities of the researchers themselves. However, certain aspects of research are mundane and repetitive. Automation holds the promise of increasing the effectiveness and efficiency of routine processes, such as scheduling, data generation, information storage, and calculations.

Quality Control

Quality control laboratories support manufacturing operations. They ensure compliance of processes and materials with predetermined specifications. Each specification depicts a desired set of values to be obtained from the testing process.

Quality assurance and quality control laboratories work closely with manufacturing organizations. They determine the suitability of materials that the enterprise buys, makes, and sells. The results of testing assure the performance and suitability of products and materials for a given use. Because of their effect on manufacturing, results must be generated in a timely manner since a whole production unit needs to act upon them. Decisions regarding thousands or millions of dollars of production capacity and materials may be held up waiting for test results.

The context diagram for a generic control laboratory is shown in Figure 4-3. In

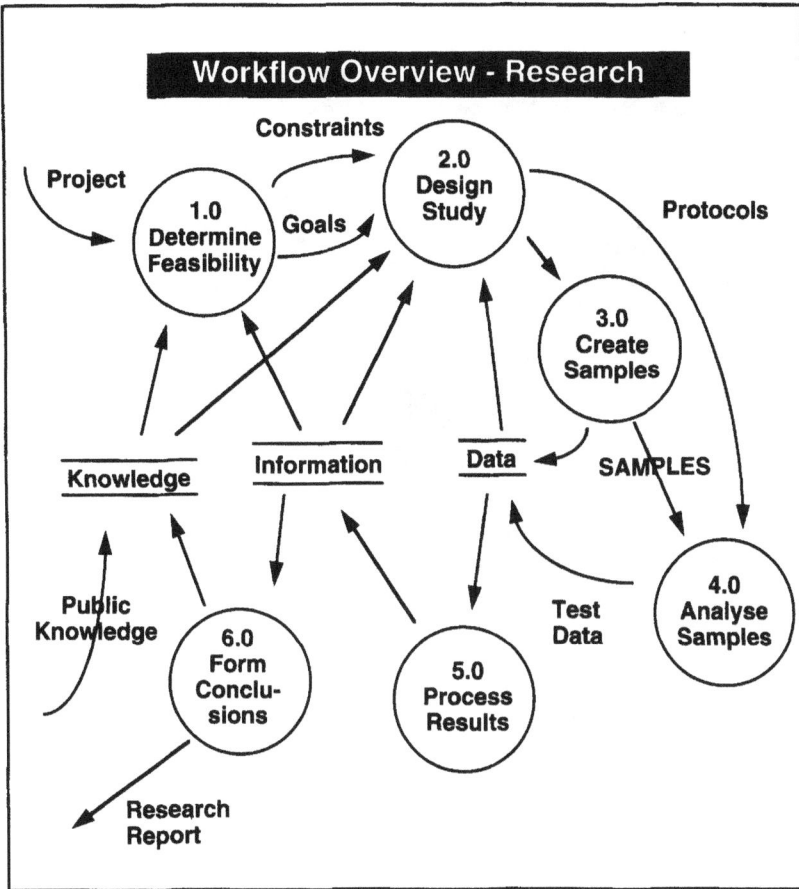

Figure 4-2 *Example of a workflow overview of a research organization*

this example, the laboratory receives samples from production. The samples to be tested may include:

- *raw materials* used as a starting point to create products;
- *intermediary materials* that result from a defined subset of manufacturing steps;
- *finished products* that have completed all stages of the manufacturing process;
- *ancillary substances* directly or indirectly related to what is being manufactured;
- *equipment* used in the manufacturing process;
- *facilities* in which the manufacturing occurs;
- *personnel* who come into contact with the materials or the equipment used.

Following testing, a disposition is determined and sent back to production who then act on the results. The control laboratory may also issue certificates of analysis to customers. These certificates document the outcome of testing completed on materials sold to each customer.

Product and process specifications may be set by an engineering group or

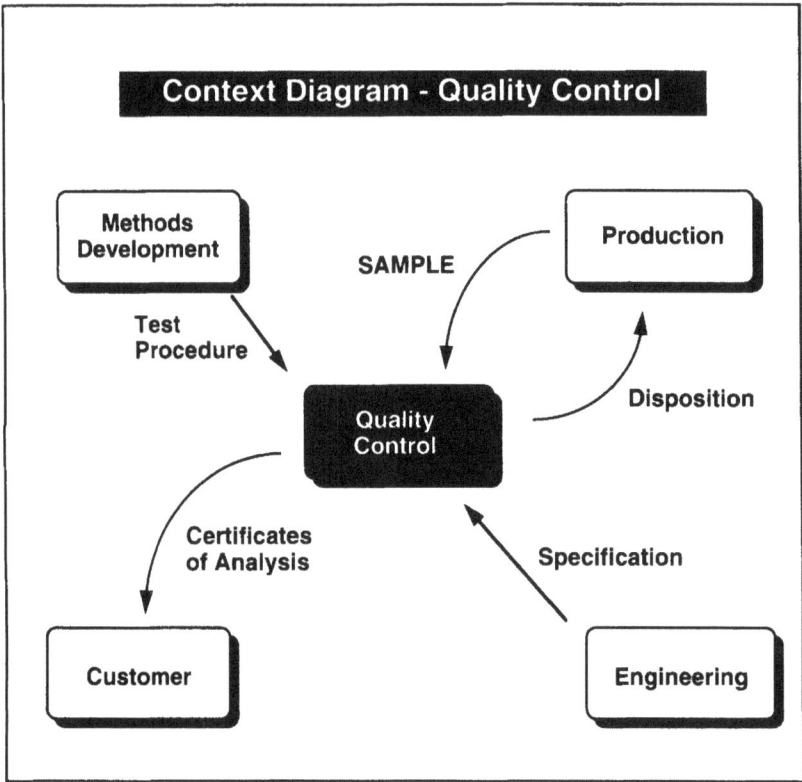

Figure 4-3 *Example of an overview of quality control*

another group outside the laboratory. The procedures for analysis may also be established by a group responsible for test methods development.

A flow diagram for a generic control laboratory is shown in Figure 4-4. Initially, descriptive information about each sample received within the laboratory is recorded.

Testing requirements are then determined. The exact analysis needed may be based on several criteria such as the product name, sample type, or manufacturing protocol. It can also be pre-determined by existing specifications. The laboratory may also elect to run additional analyses which are not covered by specifications.

Samples are then analysed. Those driven by specifications are run according to established protocols. The results of testing are then recorded.

Test results are then compared with specifications. Based on this comparison the disposition of the sample, product, or lot is communicated. The disposition determines whether the product is acceptable, should be rejected, or if another sample is required. Materials whose analytical values are within specifications are judged to be suitable for a given application. Raw materials that pass specifications can be used in the initial steps of manufacturing. Intermediary materials that pass specifications can continue with the remaining steps in the

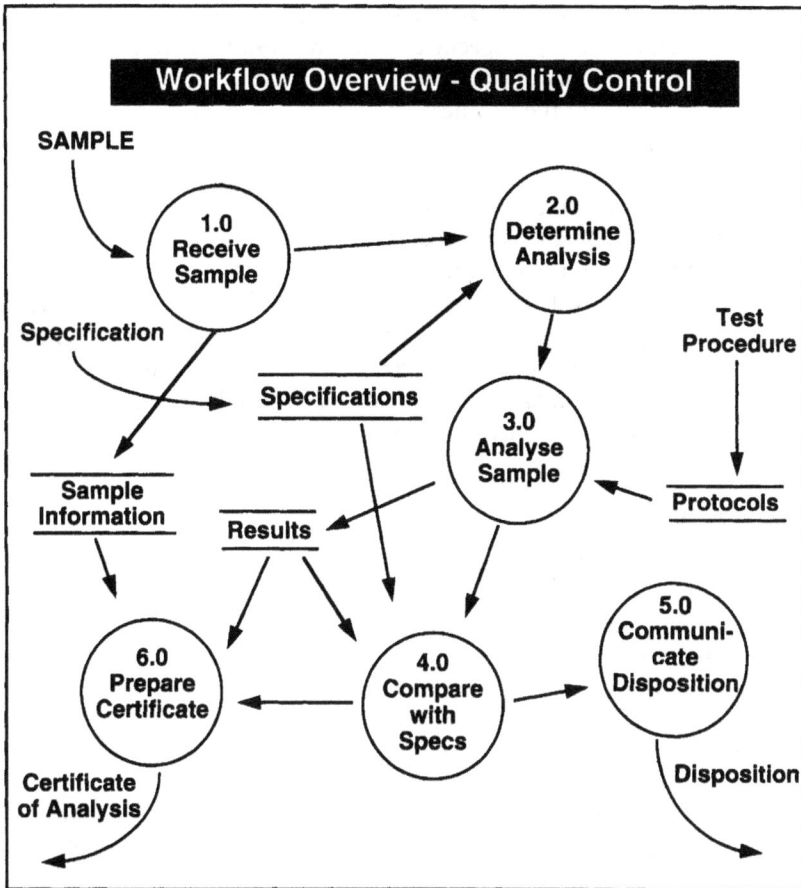

Figure 4-4 *Example of a workflow overview of quality control*

manufacturing process. Final products that pass specifications are appropriate for sale and distribution to the marketplace.

The treatment of values that are outside specifications differs significantly, even within the same organization. Materials that do not pass specifications may be rejected, remanufactured, or sold as something else with different specification limits. If test results are just marginally outside specifications, then the laboratory may repeat testing on the same sample and also test other representative samples of the material. Repetitive testing may also be performed by another analyst, using another instrument, fresh reagents, or by an independent laboratory. The intention of repeat testing is to validate the accuracy of the out of specification result. Before rejecting materials that may be worth millions, work is repeated to ensure that out of specification results really reflect properties of the materials and the manufacturing process, rather than errors in the testing process such as calculation errors or an instrument out of calibration. Organizations differ significantly in the definition and treatment of out of specification results.

Control laboratories have a large impact on manufacturing. They ensure that products are made to pre-set specifications. The results of testing ensure the performance and suitability of products for a given use. If there is a problem with a product, it is likely that the practices and operations of the laboratory will be subject to examination.

It is increasingly likely that the practices and operations of these laboratories will be audited by customers or government agencies. The soundness of the quality assurance and quality control unit dictates how well an enterprise stands up under product liability litigation or regulatory investigations. In many instances, customers insist on sound operating and information management practices on the part of their suppliers. This becomes as much as part of vendor selection criteria as the quality and costs of the product itself! In certain industries, the information handling practices of control laboratories are strictly regulated and randomly audited for conformance.[3–5]

Analytical Services

Analytical services laboratories provide testing to meet targeted client needs. Roles fulfilled include commercial testing and technical support. The exact services and analytical capabilities vary significantly. Some focus on providing testing that requires strict conformance to regulatory or professional standards. They establish their competence in specific analysis through a prolonged certification process. Others offer testing with specialized and expensive instrumentation and highly trained operators. Most focus on specific analytical techniques, industries, products, or applications.

Clients may be internal or external to the enterprise. For external customers, the laboratory maintains billing and provides discounts, especially to high-volume customers. Utilization by both internal and external clients is monitored and periodically reported. This provides a management basis on which to determine whether capabilities and services are over-utilized, or under-utilized, and whether additional capacity is needed.

Cost control is a major concern for analytical services laboratories. The magnitude of this concern is directly related to the level of competition for similar services. Each manager is constantly challenged with ensuring that his organization provides the best value for services at a competitive price. Companies faced with increased cost pressures are always evaluating options for lowering the cost of their analytical services.

A context diagram for an analytical services laboratory and its relationship to clients and regulators is shown in Figure 4-5. Clients submit samples and are provided with reports of analytical results. Clients are also billed for services

[3] 'Good Laboratory Practices; Final Rule', US Department of Health and Human Services, Food and Drug Administration, The Federal Register, 40 CFR, 1978.
[4] 'Current Good Manufacturing Practices for Finished Pharmaceuticals', US Department of Health and Human Services, Food and Drug Administration, 43:45076, 1987.
[5] 'Quality Management and Quality Assurance Standards — Part 3: Guidelines for the Application of ISO 9001 to the Development, Supply, and Maintenance of Software. DRAFT 14-February 1990', ISO, Geneva, 1990.

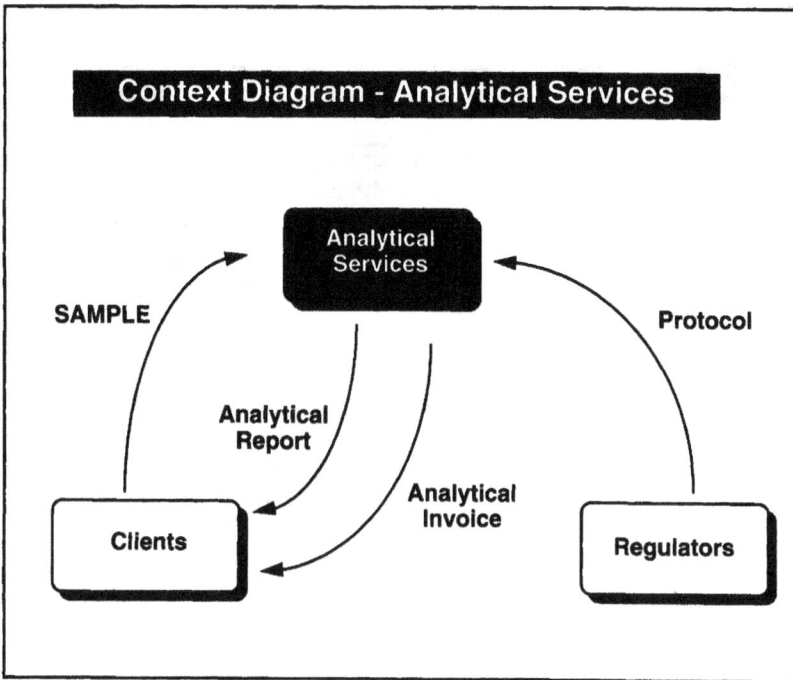

Figure 4-5 *Context of analytical services*

rendered. The regulators dictate the testing protocols used by the laboratory. They may be a government agency, a standards setting organization, or an internal group.

The overview diagram in Figure 4-6 shows the procedure following receipt of samples in the laboratory. Each sample is assigned a unique sample identifier, and information about the sample is recorded. The numbered sample is then placed in a special storage area for pending samples, which includes those that have been received but for which testing has not yet started.

Analyses are scheduled according to the attributes of each sample. For a typical service laboratory, scheduling can be based on several factors.

- Chronology. Work on samples is normally scheduled according to when they were received by the laboratory. Samples that were logged in on Monday are scheduled before those that were obtained on Wednesday.
- Priority. Samples that are designated as high priority or urgent are scheduled before those that are designated as normal samples.
- Equipment utilization. A few analyses require extensive time to set up and calibrate instruments. The same amount of time is needed to set up for one as for several samples. Laboratories will typically wait until several samples are in before conducting the given test to more effectively spread out the time used to calibrate the instrument. In this case, judgment is exercised to balance efficient

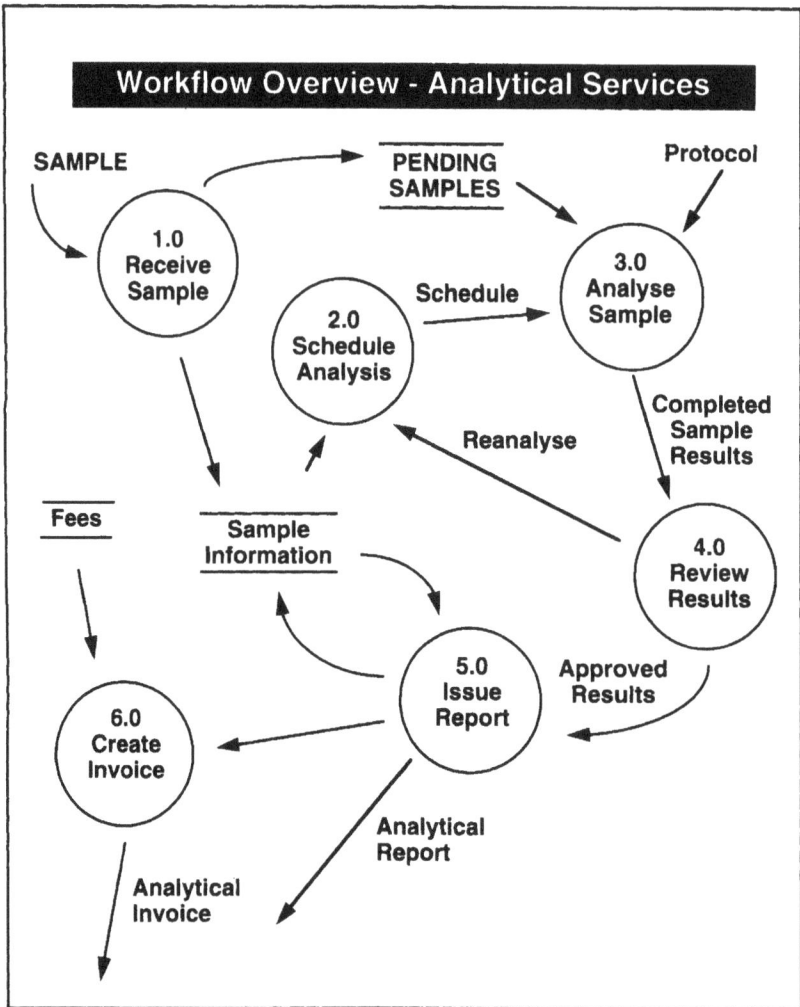

Figure 4-6 *Example of a workflow overview of analytical services*

utilization of instrumentation and personnel resources with turnaround time commitments.

Samples are then analysed as scheduled according to appropriate test protocols. Next, the results are reviewed for all completed samples and, if necessary, reanalyses are performed. Samples whose results pass this validation step are designated as 'approved results'. A report of approved results is sent to clients. The report contains information about the sample such as the sample point and submitter along with final results. Finally, invoices are issued for the work performed based on a set fee structure.

Laboratory Functions

- **Determine Feasibility**
- **Design Study**
- **Create Samples**
- **Receive Samples**
- **Determine Analysis**
- **Schedule Analysis**
- **Analyse Samples**

- **Process Results**
- **Review Results**
- **Compare with Specs**
- **Communicate Disposition**
- **Form Conclusions**
- **Issue Report**
- **Prepare Certificates**
- **Create Invoices**

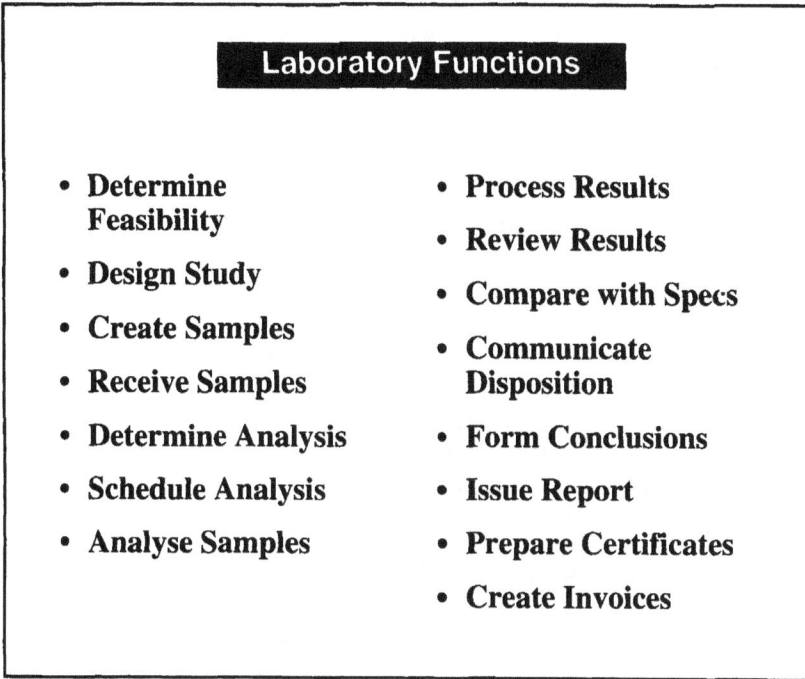

Figure 4-7 *List of laboratory functions from preceding examples*

Summary

The previous section presented simplistic representations of three different laboratory types: research, quality control, and analytical services. How similar are these models to your laboratory? I suspect that responses will range from 'close, with minor differences' at the one extreme to 'not even remotely related' at the other.

Figure 4-7 lists all the functions shown in the previously discussed models of hypothetical laboratories. This listing is by no means an exhaustive inventory of all functions performed by laboratories. While the previous discussions associated individual functions with given laboratory types, the real world dictates that any given laboratory may be composed of any combination of these functions. The exact work performed by each laboratory is driven by its unique business requirements and challenges.

Key objectives for service laboratories are to keep costs and turnaround in line with the competition. They are driven by rapid response to their customer's needs and efficient utilization of personnel and equipment. Scheduling, workload monitoring, and invoicing functions are important. Information flow revolves around samples, sample batches, and analysis.

Research laboratories are responsible for new technology. The manner in which samples are analysed and data are processed are normally not rigorously defined. Expensive resources in terms of personnel and apparatus create large amounts of data and information which are difficult to manage. Research laboratories unknowingly and needlessly duplicate a fraction of their experiments because of inadequate or no management of past results. The effect is a waste of time that could have been directed towards experiments aimed at previously untested hypotheses. Information flow in a research laboratory revolves around historical and current data as well as various projects.

Quality control laboratories are responsible for monitoring manufacturing systems for conformance with pre-determined specifications. All too frequently, quality control laboratories must not only determine if materials do or do not meet specifications, but also play a major role in determining the cause of production problems. In a quality control laboratory, information flow revolves around products, lots or batches, samples, and specifications.

There are many other laboratory environments that have not been discussed. Included are those dedicated to various clinical testing disciplines, forensics, toxicology, utilities, and various government agencies. Each operates under its own unique set of business conditions and demands.

2 Laboratories, Data, and Information

The purpose of laboratory analysis is to take physical objects and generate information about them. Each sample is representative of a larger system or object under investigation. Through a variety of analytical procedures and techniques, analysts generate test data from each sample. The data are further transformed and manipulated to determine each sample's physical, chemical, or biological properties. This analytical information assists us in making decisions or answering questions regarding our technology, commerce, and welfare.

Computer technology has introduced significant improvements in the sample handling and data collection throughputs of modern laboratory instruments. The advances are largely due to automation of sample preparation, component separations, sample introduction, data collection, and calculations. Automation has led to newer and more complex measurements, previously too time-consuming or impractical for routine applications. Examples include techniques such as Gas Chromatography/Mass Spectrometry (GC/MS), Fourier Transform Infrared Spectroscopy (FTIR), and Inductively Coupled Plasma (ICP). The newer instrument systems allow fewer analysts to generate more data in a shorter time span with smaller sample quantities.

These capabilities have greatly expanded the speed and amount of data generated from each sample. However, as stand-alone analytical systems, they have done little to improve the steps in which the data is collated and reduced into the key pieces of information upon which decisions can be made. This often requires test data from several sources in combination with information about the sample or the system from which the sample was taken.

Preceding sections of this chapter have focused on the diversity of laboratory roles and their supporting processes. This reflects only one dimension in the understanding of the inner workings of a laboratory. This section will focus on the complex web of interactions between laboratories, data, and information. Understanding these intricacies is important because it impacts the selection and design of your LIMS. The topics and issues presented in this section will hopefully provide you with a starting point for these considerations.

Dimensions of Laboratory Data

McDowall and Mattes,[6] Liscouski,[7] and Mahaffey[8] present a hierarchical segmentation of information resources within the laboratory as data, information, and knowledge. Their breakdown is consistent with works on information resource management.[9,10]

The segmentation of data, information, and knowledge is as follows.

- *Data* consists of raw facts that are either acquired or observed about people, objects, or events. Examples of analytical data include values directly obtained for mass, volume, and detector response, observations, spectra, and chromatograms.
- *Information* is data that are manipulated and presented in a form suitable for human interpretation and decision making. It includes facts such as concentration levels or substances present in the sample.
- *Knowledge* are facts uncovered from the information. It tells us that a certain formulation will or will not work in a given process. It assists in determining the direction of future research efforts. Examples of knowledge obtained from laboratory analyses are, the 'water is safe to drink', 'our cat is healthy', 'material from today's production run is suitable for release', and 'the hypothesis underlying the current research project is not valid'.

Raw Data

Raw data is the *original record* of a measurement or observation. Data entered into a computer directly via the keyboard or automatically from an automated instrument is considered to be raw data. However, if the data is transcribed to the computer from a source document, such as a laboratory worksheet, notebook, or instrument printout, it is not considered to be raw. In these instances, the source document is the raw data.

It also includes original descriptive information provided about the sample and its source. This may include a history of the sampling, a listing of materials included, or other attributes of the sample submitted for testing.

Data that is in any way processed is no longer considered to be raw. Raw data precedes any subsequent calculation, manipulation, interpretation, modification,

[6] R. D. McDowall and D. C. Mattes, *Anal. Chem.*, 1990, **62**, 1069A.
[7] J. G. Liscouski, *Anal. Chem.*, 1988, **60**, 95A.
[8] R. R. Mahaffey, 'LIMS Applied Information Technology for the Laboratory', Van Nostrand Reinhold, New York, 1990.
[9] F. R. McFadden and J. A. Hoffer, 'Database Management', Benjamin/Cummings, Redwood City, CA, 1991.
[10] D. W. DeHays, J. A. Hoffer, E. W. Martin, and W. C. Perkins, 'MIS for Managers', MacMillan, New York, 1991.

or any other form of processing. They serve as a basis upon which all aspects of the original work can be re-derived, reconstructed, and reinterpreted by competent experts.

In practice, what is considered raw data can differ significantly.

- Many computerized instruments including chromatographs, spectrophotometers, and pH meters provide final concentration values based on preliminary calibrations. The values displayed are calculated by programs within the instrument itself. The original, unprocessed measurement may not be readily available to the analyst. In this event, the values given directly by the instrument and the accompanying calibration records are considered to be the raw data.
- In certain instances, the exact analytical conditions and protocols are subject to high levels of variability and uncertainty. Analysts may proceed through a number of preliminary tests and generate large amounts of data in an attempt to obtain a valid analysis. In this event, only the data associated with the scientist's professional determination of what constitutes a valid analysis is further processed and interpreted to yield information about the sample and its properties. Data from the invalid analysis is not used.
- In other instances, all work is retained as raw data. The conditions under which the work was completed and the data obtained from both valid and invalid analyses are retained and interpreted by scientists.

Processed Data

Processed data is an outcome of a calculation, manipulation, or other processing of the original raw data. Examples of processed data include a baseline-subtracted chromatogram, a statistical average of several points, or a listing of retention times and peak areas from a chromatogram.

Intermediate Results

Intermediate results have units or context relative to the properties determined by the analysis. Examples include component concentrations, substances present, or organisms identified. Intermediate results are associated with a single property and one instance of a test. Examples include the outcome of testing on individual replicates, sample aliquots, or discrete samples and resamples.

Final Results

Final results are the final outcome of one or more analyses for an analytical property. It represents the actual values or characteristics of the sample determined from all tests performed. Final results may be rounded or otherwise qualified to reflect the accuracy and precision of the analysis. It may also reflect a statistical calculation of intermediate results from several samples, replicates, or aliquots.

Reported Data

Reported data include all information released from the laboratory and directly applied to an evaluation or decision. They may contain final results which have been summarized, averaged, plotted, or otherwise manipulated to facilitate and expedite the decision-making process. Reported results may also contain intermediary results, processed data, or any other data that supports the decision-making process. This may include details from the analysis such as chromatograms, photographs, or spectra.

Exactly what is released by the laboratory differs significantly. In some cases, the laboratory only provides raw data to its clients. Groups or individuals outside the laboratory process and interpret the raw data. An example may be a spectroscopy laboratory servicing a group of synthetic chemists. At the other extreme, the laboratory may be intimately involved with critical decisions based on analytical results. In this case, the laboratory may issue only its conclusions and findings.

Manifestations of Laboratory Data

Media

Laboratory data is manifested by an assortment of physical media such as:

- microfilm;
- microfiche;
- computer printouts;
- magnetic or optical computer storage media;
- recorded data from automated instruments;
- recorded data from mechanical devices (strip chart recorders);
- laboratory worksheets;
- laboratory notebooks;
- paper records;
- memoranda;
- notes;
- photographs;
- video;
- audio.

Formats

Laboratory data can take the basic forms of images, text, numbers, and electronic output. Images include photographs, drawings, and video. Photographs and drawings portray an image of an object at a fixed point in time, whereas video depicts images over a range of time. Images also include graphic plots of spectra and chromatograms. Text consists of short descriptive statements of a few words to long multi-page documents. Numbers represent quantitative values for specific properties or measurements. To be meaningful, numbers must be accompanied by their associated units and the property quantified. Electronic output include recordings of sounds, chromatograms, and spectra. These outputs can be reconstructed, replotted, and otherwise manipulated.

Challenges

An integrated approach to the management of all the various physical manifestations and types of laboratory data has been elusive. Specialized automated systems handle one type of media or one type of data well. Most high-performance database management systems of the early 1990s are well suited for numeric and short textual data input through a keyboard. Handling the totality of other laboratory data involves separate systems and technologies. Bringing it all together poses large management and technical challenges.

Utilization of Laboratory Data

Laboratory data can be further segmented according to the way in which it is used. Operating data meet short-term, transactionally oriented needs. Strategic data fulfil long-term needs.

Operating

Operating data supports the daily business activities of the laboratory. This includes the detailed information necessary to complete testing on samples and report results. Speed is an important characteristic of systems designed to handle operating data since it affects the daily performance of the organization. Operating databases include a lot of detailed information about the outstanding work within the laboratory.

Strategic

Strategic data are typically subsets or summaries of the operating data. They include information necessary for management planning and control such as summaries of the laboratory's workload, testing demands, and performance. Strategic data also include the key information for long-term trending and statistical analysis of test results. Strategic databases include long-term summaries of work already completed by the laboratory.

CHAPTER 5

Laboratory Interactions with Other Groups

The nature of laboratory information and its corresponding processes are affected by interactions with parties outside the laboratory organization itself. A laboratory's exact relationship with external groups varies considerably. Groups outside its direct control are frequently responsible for tasks that directly affect the laboratory. These include activities such as providing samples, establishing test procedures, and maintaining specifications. Conversely, the laboratory also affects the jobs of others, especially those who need to make decisions based on test results. While the individuals responsible for these tasks may not be part of the laboratory, the manner and competency in which these jobs are done directly impacts the overall quality of laboratory operations.

The laboratory is but one of may cogs in a wheel that plays a major role in the inventing, making, and servicing of the organization's products. The LIMS can facilitate or impede the smooth execution of critical processes. The potential benefits of automation may not be achievable if these external interactions are not considered as the system is implemented.

In some instances, the compelling reasons for the LIMS are benefits outside the laboratory. For example, implementation in a clinical setting may be driven by improving the dissemination of patient results, thereby improving the speed in which physicians can initiate treatments. While laboratory efficiency may improve, the major driving force for the LIMS, in this particular case, rests with improving patient care.

With increasing frequency, laboratories are being asked to not only provide test results, but also actively participate in key decisions based on those results. To do so the LIMS may need to include additional databases to track information about the organization's products, services, projects, and customers. Instead of just testing samples to generate results, laboratories are increasingly expected to explain the context of analytical results relative to the original source or process from which the samples were taken.

Implementation of the LIMS provides an opportunity to improve how processes operate not only within a laboratory, but also for other organizations that rely on laboratory data. Why automate processes that are inefficient, that poorly meets needs, or that do not work well? It is important to focus on those who may be affected by the LIMS and not to restrict your view to the laboratory

63

only. This includes the totality of needs including: that of the laboratory, the enterprise to which it belongs, and the customers served. In todays rapidly changing and leaner organizations, jobs that are today completed outside the organization targeted for the LIMS may, in the near future, be added to its list of responsibilities.

1 Groups Within the Enterprise

A context diagram, as described in Chapter 3, graphically models the boundaries of laboratory operations. It serves as an important tool to depict the various groups that the laboratory must interact with. Examples include groups such as Manufacturing, Analytical Methods Development, Quality Assurance, Process Engineering, Accounting, and Regulatory Affairs.

Once the exact groups have been identified, it is then necessary to determine the role played by each external group. 'Role' describes a specific functional responsibility within the organization. A given role may be assigned to one or more groups. Conversely, any given group may be responsible for more than one role. Role names should not be confused with the names of organizational units. Roles define functions. Organizations define related groups of people within the enterprise. The names and functions performed by groups within organizations change frequently; the roles or functions performed within the enterprise generally do not. The actual names given to groups and their assigned roles differ significantly from one organization to another.

The context diagram for the laboratory may bear little resemblance to the context diagram for the LIMS. Some of the roles fulfilled by groups outside the laboratory may actually fall within the scope of functions automated by the LIMS. Conversely, specific functions typically carried out by the laboratory may remain outside the scope of the LIMS. They may be handled manually or become automated by other systems.

The remaining paragraphs of this section present various roles typically fulfilled by groups outside the laboratory, how they affect operations of the laboratory, and their potential impact on the LIMS.

Sample Provider

The sample provider is responsible for the acquisition, packaging, labelling, and transportation of samples to the laboratory. Specialized protocols may be needed for the handling of samples prior to analysis. If they are not followed the technical validity of results obtained from testing may be compromised. Sample providers are also responsible for proper identification of the sample and its source. Improper identity of the sample jeopardizes the soundness of decisions based on analytical results.

Ways in which the LIMS can assist sample providers include:

- maintenance of sampling schedules and periodic creation of sampling lists;
- creation of container labels to uniquely identify samples and their source;

- generation of appropriate sampling, packaging, storage, and transportation instructions to ensure that appropriate protocols are followed.

The answers given to the following questions determine how sample providers can affect the LIMS.

- How should samples that were improperly gathered be handled? Should the laboratory reject the sample entirely? Should it proceed with testing and qualify the results? Should it demand a resample?
- Does the LIMS need to reconcile disparities between the samples scheduled to be taken with those actually gathered and sent to the laboratory?

Specification Provider

Specification providers determine the test methods and acceptable result values that should be applied to the routine analysis of products and other materials. Specifications determine if the products, processes, or materials tested are suitable for a particular use.

The LIMS can assist specification providers by:

- ensuring that any new or revised specifications are uniformly used by all laboratories on the LIMS;
- maintaining a base of historical results to determine if specification changes are called for;
- providing statistical and graphical trends of products and processes.

Specification providers can also affect the LIMS as follows.

- Special considerations may be necessary in cases where multiple specification limits apply to a material. Examples include separate specifications for different countries, product applications, or whether or not the material will be used in regulated or non-regulated settings.
- Are special processes required for results that are out of specifications? In certain cases, the LIMS automatically schedules retesting or resampling for results that fall outside the prescribed limits. Are the same tests repeated or are other tests assigned?
- Numerical values that are close to the specification limits pose special problems. Such values may be within or outside specifications, depending on the number of significant digits and rounding algorithms which are applied. LIMS processes can automate the comparison of test results with specifications. However, those responsible for setting specifications need to carefully define the appropriate handling of significant digits and the rounding of numerical values.

Method Provider

Method providers establish standards for the procedures and equipment required for each test. Each method determines the appropriate protocol for obtaining analytical results on a specific property or characteristic of the sample. Methods standardize the process through which test results are obtained. They ensure that result values across multiple samples analysed at different points in time are comparable.

Methods eliminate test result value variations arising from differences in equipment, reagents, analysts, and time.

The LIMS can assist method providers by:

- standardizing the way that analytical data is captured and processed;
- associating each analytical result with the method or method version used to obtain it;
- enforcing consistent and uniform laboratory-wide transitions to new or revised methods;
- ensuring that calculations are consistently executed and applied;
- maintaining a base of historical results for the evaluation of method performance and to determine if revisions are warranted.

Method providers can affect the LIMS. The methods currently in use may lack the necessary detail to capitalize on the data checking and automation capabilities of the LIMS. Items such as calculation formulae, units, and definition of the exact data to be recorded for each test may not be included within the organization's methods. In some cases, the actual properties to be reported as final results are not established. These details are left to the discretion of each laboratory group or analyst. A great deal of effort may be required to define the exact data requirements of analytical methods before you can fully take advantage of the LIMS.

Study manager

Study managers develop the requirements for scheduling, sampling, and tests to run on samples. They are responsible for the development and execution of specific studies or programs. The degree to which studies are supported by laboratory testing varies considerably. Study managers collate, maintain, and analyse all relevant study data. Testing services are requested both on a regularly scheduled as well as on an *ad hoc* basis.

The LIMS facilitates the work of study managers by:

- providing mechanisms to track the status of testing in the laboratory;
- ensuring consistent execution of study-specific testing protocols by the laboratory;
- maintaining a base of historical analytical results for each study.

Study managers can also affect the LIMS as follows.

- If study managers are granted access to the LIMS, mechanisms may be required to limit their access to data and functions related only to studies for which they are responsible. It may be necessary to restrict access to information related to other studies and other work performed within the LIMS.
- In some cases, studies generate a high volume of samples and tests. Programs may be required to expedite or eliminate the need to manually repeat the entry

of information that is identical over a large group of samples. Information is entered once, and replicated over all samples to which it applies. Such a provision would reduce transcription errors and improve overall analytical turnaround.

Result Analyser

Result analysers receive, manage, and interpret the results of analytical testing. They make decisions based on information from several sources, including the laboratory. Result analysers may study statistical or other trends across studies, materials, products, and processes.

The LIMS can assist result analysers by:

- providing statistical or other data analysis tools;
- providing linkages to access other sources of information such as commercial databases, news retrieval, and scientific literature services.

Result analysers can affect the LIMS. Mechanisms may be needed to grant access to data of specific interest to each result analyser while restricting access to other data on the system.

Other Laboratories

The laboratory targeted for the LIMS may provide specialized testing services directly to other laboratories. Conversely, the laboratory may subcontract specific analyses to others. If the volume of interactions with other laboratories is high, then it may be worthwhile considering electronic linkages to facilitate the exchange of information about samples, test requests, and analytical results.

Cost Accounting

Cost accounting is responsible for periodically creating breakdowns of laboratory charges, either for internal management monitoring or for direct billing to customers. The LIMS can assist by providing total counts of samples submitted, tests requested, and tests completed during a defined time period. A more detailed breakdown is possible only if the key elements are associated with all work processed by the LIMS. This includes elements such as submitter, charge number, study identifier, project, or responsible unit.

The needs of cost accounting can affect the LIMS as follows.

- Are charges to be based on standard test times or by actual time spent in analysis? Is there an existing base for standard test times? How will actual test times be input into the system?
- Is there a requirement for a breakdown of charges to each analyst? If so, how will the system be aware of tests that involve more than one analyst?

Product Development

Product development is responsible for assessing the commercial merits of targeted compounds, materials, or processes. They determine if such improvements are feasible, if they address a need, and if further refinement is warranted. They rely on laboratory testing to ascertain characteristics of the candidate product and to determine the measurable effects of its use. Product development often includes the previously discussed roles of sample provider, study manager, and result analyser.

Process Engineering

Process engineering specifies the procedures, facilities, equipment, and expertise needed to implement routine production of new or improved products. Analytical testing provides supporting data for the conversion of technology from laboratory to full-scale commercial manufacturing. The types of samples tested as well as the applicable specifications and methods evolve during this process. Process engineering often includes the previously discussed roles of sample provider, study manager, and result analyser.

Product Manufacturing

Product manufacturing is responsible for the repetitive creation of products for distribution into the marketplace. They are responsible for all steps in the manufacturing process. This includes initial acceptance of raw materials and the totality of steps leading to the creation of approved final products. Within an organization, the various responsibilities of product manufacturing are generally distributed among several units such as quality assurance, manufacturing, quality control, inventory management, and materials management. The actual names used and the distribution of responsibilities varies from one organization to another.

Interface Groups

Selected individuals or groups are responsible for serving as intermediaries with parties outside the organization. They coordinate the work of numerous groups within the organization to ensure coherence in dealings with external groups. The interface groups serve as a focal point of interaction with all external parties, including other businesses and government agencies.

Relative to the laboratory, these interface groups may coordinate the submission of external samples to the laboratory. They may also summarize, analyse, and report on conclusions reached as an outcome of analytical testing.

The LIMS can assist the work of these interface groups in cases where the handling of laboratory data is well defined. The system may be able to format analytical results to facilitate their use by the interface groups. In some cases, reports may be generated in a form that eliminates the need for reformatting and retyping. Other measures may be considered to facilitate the analysis and manipulation of analytical data by these interface groups. Such efforts may improve the efficiency and performance of the organizations dealings with external groups.

Examples of interface groups are those that deal with:

- regulatory agencies;
- suppliers;
- customers.

2 Groups Outside the Enterprise

With increasing frequency, the LIMS must do much more than track testing and collect results from samples. The information and operations of the laboratory are frequently an integral part of transactions between the enterprise and both its customers and suppliers.

Customers

External customers are the firms who purchase the goods or services of the enterprise. Laboratory data is critical in the creation and support of the various products sold to the marketplace.

Customers of Manufacturing Organizations

Manufacturing organizations may provide customers with certificates of analysis. Each contains the analytical results obtained from the actual batches or lots of materials purchased. The results are accompanied by either a handwritten or electronic signature. The certificate assures customers that the materials have been tested and that the results fall within acceptable limits.

The handling of certificates of analysis differs considerably from one organization to another.

- The actual testing is usually performed just prior to completion of the manufacturing process. However, where the material is subject to prolonged storage or when it changes rapidly with time, the analysis may be performed just prior to shipment.
- The analytical results reported usually represent a subset of all testing performed during the entire manufacturing process. Results from the various starting materials and product intermediates are not normally included. Even certain tests on the finished product may be excluded. For example, the analysis of proprietary substances or additives may be excluded from a certificate of analysis. Inclusion of such information may disclose trade secrets regarding either the manufacture or composition of the organization's products. Mechanisms must be in place for the LIMS to differentiate tested properties that should be externally disclosed from those that should not.
- The exact information presented on the certificates of analysis may differ from one customer to another. Certain customers may request specialized testing or

formatting for their certificates. This may eliminate the customer's own need to perform testing or it may expedite their internal processing. Organizations are generally receptive to these special requests if the appropriate financial incentives are in place. In this case, the LIMS needs to differentiate customer-specific certificates.

- Some customers demand electronic transmission of certificates of analysis by Electronic Data Interchange or other direct computer to computer linkages. These arrangements serve to hasten the movement of information and the processing of transactions between customers and suppliers. They also serve to reduce the effort necessary to handle and distribute paperwork.

Customers of Service Organizations

Service organizations rely on laboratory testing to assess existing conditions, develop service programs, and determine the effectiveness of services provided. Analytical testing may only support the principal service provided by the organization. Examples include testing which supports medical care, engineering design, and feasibility studies. In other instances, such as commercial testing laboratories, analytical reports are the primary product of the organization.

In a service setting, laboratories need to fulfil highly focused needs consistent with the primary services provided by the organization. The manner in which they operate is highly dependent on the nature of services provided, the technical disciplines involved, and the existence of any applicable regulations or industry standards. The LIMS needs to focus on the specific functional, information, and regulatory requirements that apply to the organization and its services.

Suppliers

Suppliers provide materials used in the creation of the organization's products or services. In some cases, the materials must conform to established specifications before they can be used. Analytical testing of the materials may be done by the laboratory. In some cases, suppliers provide a certificate of analysis as evidence that their materials are suitable. The laboratory may confirm the suppliers test results by repeating all tests or a select subset of tests. It may be necessary to relate testing data provided by suppliers to analytical results within the laboratory.

Standard Bearers

External standard bearers establish uniformity in practices carried out by multiple organizations. The standards facilitate information sharing and teamwork. Standards of potential concern to a LIMS include:

- standards for the documentation of procedures and responsibilities for the implementation and operation of both manual and automated laboratory information systems;[1]

[1] J. B. Doherty, *Chemometrics Intelligent Lab. Syst.: Lab. Inf. Manage.*, 1991, **13**, 135.

- standards for analytical methods utilized by laboratories;
- standards for electronic analytical data exchange.

3 Regulatory Compliance

Many aspects of commerce are constrained by government or outside agencies. This is especially true of activities that have a direct or indirect bearing on human health and welfare. Typical examples of the industries affected are pharmaceuticals, nuclear, foods, aerospace, and chemicals. Participation in the regulated marketplace has several characteristics.

- Laboratory testing is required to substantiate claims of quality, safety, and effectiveness.
- Test results must be disclosed to external agencies.
- The test data are subject to extensive external review and analysis.
- The business practices surrounding the data creation and use are subject to periodic challenges to ensure that they are effectively managed and consistently applied.

Regulations were first introduced in the late 1970s in response to deficiencies uncovered through investigations of testing laboratories. The investigations found poor data management practices which compromised the overall credibility and integrity of laboratory data. Evidence of malicious data fabrication and fraud was also uncovered.

Guidelines and regulations were established to restore confidence in the quality and integrity of data upon which critical decisions are based. They are embodied in Good Laboratory Practices (GLPs)[2] and Good Manufacturing Practices (GMPs)[3] initially established by the US Food and Drug Administration. Acceptance of these principles has gradually spread into other regulatory groups, industries, and countries.

Unfortunately, the guidelines were developed prior to the routine use of laboratory computers. They primarily focused on manual operations. However, regulators claim that the dominant issues for manual information management also apply to automated systems. To conform the laboratory needs to provide:

- evidence that reasonable mechanisms are in place to protect data from corruption or destruction (both intentional and unintentional);
- inspections to ensure that calculation formulae and processing algorithms are accurate and appropriately applied;
- mechanisms to ensure that data entry and other functions can only be made by authorized individuals;
- the retention of all data entered and any subsequent changes to the original data;
- mechanisms to track the entry and changes to all data to a responsible person at a fixed point in time;

[2] 'Good Laboratory Practices; Final Rule', US Department of Health and Human Services, Food and Drug Administration, The Federal Register, 40 CFR, 1978.
[3] 'Current Good Manufacturing Practices for Finished Pharmaceuticals', US Department of Health and Human Services, Food and Drug Administration, 43:45076, 1987.

- evidence that personnel are appropriately trained and qualified to perform their assigned responsibilities;
- archives for the orderly storage and expedient retrieval of all raw data, documentation, protocols, specimens, and reports;
- operating and maintenance procedures which are clearly documented and easily followed.

The GLP guidelines are notably lacking in specifics regarding the automated handling of laboratory data. Many of the terms lack clear definitions and they are open to a wide range of interpretations. Organizations seeking compliance have been frustrated by unsuccessful attempts to obtain unambiguous guidance from the regulators. They are generally unable to get definitive responses to specific questions. The absence of definitive standards leaves both laboratories and auditors with little guidance in instituting measures to meet regulatory expectations.

With the increasing use of computers in laboratories during the late 1980s, regulators directed increased attention to guidelines for Good Automated Laboratory Practices (GALPs). This extension of GLP includes more detail on the procedures, responsibilities, management, and use of computers for the handling of laboratory data.

The US Environmental Protection Agency (EPA) has been particularly active in their pursuit of GALP guidelines.[4] They rely heavily on laboratory testing to make decisions on public health and environmental issues. Most of the analyses are completed by contracted laboratories outside the direct control of the EPA. The agency realized that most laboratories were replacing manual operations with computer technology. Their policies and decisions increasingly relied on analytical data from automated systems. Findings of fraud, possible corruption, loss, and inappropriate modifications of computerized data underscored a need for definitive agency-wide standards.

This need was confirmed by a 1990 study of several automated laboratories and their practices. The agency concluded that the integrity of computer-resident data is at risk in many laboratories providing scientific and technical data to the EPA. Serious gaps in system security, data validation, and documentation are responsible for this risk. Major areas of non-compliance have also been noted in numerous inspections conducted by the Food and Drug Administration.

A few of the specific problems encountered by the agency and others were as follows.

- Testing dates were manipulated for samples that were to be analysed within prescribed time periods. This was done through adjustments to the internal clock of the instrument, or by changing values for data already stored on the computer.
- Instrument calibrations were electronically manipulated or other calibrations used in cases where the actual test results did not meet range specifications.

[4] 'Good Automated Laboratory Practices (Draft): Recommendations for Ensuring Data Integrity in Automated Laboratory Operations with Implementation Guidance', US Environmental Protection Agency, 1991.

- Computer data was manipulated to report results on samples that were never analysed.
- Results indicating non-compliance with regulatory specifications were withheld or destroyed.

Most problems came about from workers acting without the direct consent or knowledge of management. However, the organization as a whole, and the specific mangers of the units, were held accountable for these violations. They may not have been directly involved or have even endorsed the actions of their subordinates, but they are accountable for not having the appropriate controls and processes in place to protect the integrity of data produced by their organization.

The major areas of automated system problems and deficiencies generally noted in laboratories have been in the areas of:

- system security;
- data validation;
- data change practices;
- system documentation;
- computer system validation;
- quality assurance.

LIMS Considerations

This section discusses how compliance with regulatory guidelines affects the implementation and operation of a LIMS. The effects are notable in the area of the system's reliability, activities, and costs.

Reliability of Laboratory Data

Compliance with GLPs and GALPs provides external groups with confidence that sound decisions can be based on laboratory results. It creates tangible evidence of the laboratory's commitment to consistently producing technically credible and high-quality data. They include processes to ensure that:

- consistent practices are followed for the creation of laboratory data and reported test results;
- all data are critically reviewed and, if necessary, corrective actions are implemented before any test result is reported;
- all data that are created, manipulated, and approved are directly traceable to a responsible person, system, or program at a set point in time;
- all original data and any subsequent changes are retrievable;
- reasons for all data modifications are recorded;
- reported results can be substantiated through the retrieval of their supporting raw data and the reconstruction of all subsequent events and processes;
- capabilities for the entry, modification, and approval of data is restricted to authorized and appropriately trained personnel;
- reasonable measures are in place to prevent both the intentional and

unintentional loss or corruption of data arising from power failures, system failures, operator errors, and malicious intent.

Reliability of LIMS Operations

Compliance ensures external groups that the LIMS has, is, and will continue to operate in a reliable and predictable manner. To meet this goal, operations of the LIMS should be consistent, traceable, robust, and controlled. This includes documented management processes and evidence substantiating that:

- clearly identified and properly qualified individuals are responsible for the development, operations, management, use, and maintenance of the LIMS;
- the LIMS is used and maintained in a consistent manner as described in clearly written and management approved Standard Operating Procedures (SOPs);
- deviations from the SOPs are recorded and reviewed;
- all tasks related to problem resolution, maintenance, configuration modification, and SOP approval are traceable to a responsible individual at a given point in time;
- calculations and other automated processes are working properly;
- reasonable measures are in place to recover operations in the event of personnel turnover, disaster, and malicious intent;
- changes to the system or procedures are carefully reviewed for their impact on the operation and reliability of the system.

Activities

Regulatory compliance includes measures to ensure that the system is installed, used, and maintained in a manner that ensures proper and reliable operations. To some degree, they should be considered for the implementation and use of any automated system, whether or not regulatory guidelines apply. The actual measures adopted in a non-regulated environment should be based on the criticality of the system, the impact of decisions based on its results, the potential liabilities of corrupt data, and the degree to which the data is subject to audits by groups outside the enterprise.

Regulated applications generally require much higher amounts of documentation and management controls. This means that more effort is required to:

- create documentation that is detailed, clear, and understandable by external auditors who are generally unfamiliar with the terminology and operations of the organization;
- establish controls for the creation, review, approval, storage, revision, and distribution of procedures and other documentation;
- provide for independent inspections and audits of the development, operation, and maintenance of the system;
- maintain detailed records of system maintenance;
- control and record all changes to the systems hardware and programs;
- provide evidence of orderly evolution of the development, installation, and use of the system.

CHAPTER 6

Impact of a LIMS on the Laboratory

A LIMS drastically alters the handling of laboratory information. The new system can improve the performance, efficiency, and capabilities of the laboratory. It can also result in unanticipated effects, which if not handled appropriately and expediently, will have undesirable effects on the organization's performance. The magnitude and impact of these effects is largely determined by the scope of implementation. Factors such as the number of functions to be automated, the number of people using the system, and the volume of data (or tests) processed determine the magnitude of the LIMS's affect on the laboratory. The positive and negative impacts of automation will be less keenly felt for implementations that are less ambitious.

Working with the new system alters previously accepted paradigms regarding operations and management of the laboratory. As with most automated systems, a LIMS changes the way that the laboratory and individuals do their work. A discussion of how the organization manages the numerous changes is presented in further detail in Chapter 14. The remainder of this chapter highlights the effects of automation on the laboratory as a whole and on the individuals who work within laboratories. This is followed by an overview of the common misconceptions of a LIMS's impact on the laboratory.

1 Effect on Individuals

The automated technology alters the normal working routine of individuals. Working with computers has significant ergonomic and procedural differences from manual information management practices based on pencil and paper. Shoshana Zuboff has studied the effects of automation on the working life of individuals.[1] The net effect is an abstraction of work, diminished direct people to people interaction, a need to develop new skills, and higher levels of attention required in the workplace.

[1] S. Zuboff, 'In the Age of the Smart Machine: The Future of Work and Power', Basic Books, New York, 1988.

Diminished Contact with Work Products

With the introduction of automation, workers come to feel physically separated from the tangible results of their individual work efforts. The LIMS alters previously accepted ways for individuals to obtain a sense of their own accomplishments. It also imposes a physical separation from the information. From a user's perspective, information in a computer mysteriously resides in a database connected to the other end of the terminal, rather than on pieces of paper which can be seen and felt. In a manual system, progress and the amount of work is readily apparent by visual or other sensory clues, such as the height of the paper stacked on the corner of the desk. With automation, feedback on the quantity of work must be obtained through the LIMS database; information which may or may not be conveniently available to everyone in the laboratory. Features provided by the new system need to either replace or compensate for the various sensory elements eliminated by automation.

Altering Frequency and Nature of Direct Interaction with Co-workers

Information stored in a centralized database is shared without the movement of paper, telephone calls, or a variety of other means requiring direct interaction with another person. In cases where the LIMS serves as the nexus of the laboratory's information management needs, the frequency and intensity of personal interactions with co-workers and supervisors diminish. The net effect may be an overall operational improvement, but the change drastically alters person to person alliances and working relationships within the organization. Some workers may acquire a feeling of isolation in their working community, resulting in lowered morale and non-productive attitudes. Others may see the change as an improvement by reducing the need for non-productive or negative interactions with others.

Higher Levels of Attention

Completing daily work assignments through the LIMS requires the input of information through screens, or a series of screens, on the system. An individual's speed is dictated by the mastery of skills in operation of the various system programs. These skills and the speed at which individual operations are accomplished increase with prolonged use of the system. However, factors affecting performance of the computer system often result in unpredictable and dramatic slowdowns in the operation of programs. The speed with which programs navigate from one screen to another changes according to demands placed by other users and other activities on the system. In such a multi-user environment, an individual's pace of work is dictated by factors which are beyond their direct control. The pace of work fluctuates with the ever-changing response time and performance of the computer. This requires higher levels of attention to complete even the most routine and repetitive tasks. In a manual system, the pace and rhythm of discrete tasks are directly set by each individual.

New Skills Required

The manner in which tasks are completed are drastically altered by the introduction of the LIMS. The replacement of manual procedures through automation alters the base of skills needed to fulfil daily operating needs. Many of the older procedures, and the skills necessary to execute them, are rendered obsolete and replaced by tasks completed through the new system. Individuals working within the laboratory need to acquire new skills to deal with the changes and new capabilities of the system.

2 Effects on the Organization

The LIMS will have a substantial effect on the laboratory and the various organizations served by the laboratory. The technology holds the promise of improved efficiencies, increased quality of services, and the introduction of new capabilities enabled by the automation.

The potential uses of the LIMS are not generally realized at the outset. Creative and strategic application of the capabilities are identified following a period of prolonged use and assimilation into the organization.

The potential benefits of a LIMS are discussed in further detail in Chapter 12. This section summarizes the effects of the LIMS on the organization's processes.

Standardization of Processes

How laboratory procedures are implemented by the LIMS software formalizes each laboratory's assumptions regarding the detailed steps of how each process should be performed. Through the LIMS software, selected laboratory procedures are routinized and consistently executed. The exact level of discretion permitted for each task is limited by how the LIMS software is configured for a particular laboratory environment or testing activity.

In many cases, the LIMS introduces a standardization of procedures that were previously undefined or inconsistently applied. Many laboratory information handling practices and testing procedures remain undocumented and the details of their execution varies considerably from one individual to another. Some that are documented provide only vague guidelines whose exact execution is totally reliant on the judgment, training, and experience of the individual. Automation by a LIMS frequently forces the organization to articulate laboratory needs and processes. Where these attempts are successful, the result is an overall reduction in ambiguity and improved consistency in laboratory practices.

In other situations, laboratory processes are extremely complex, they defy rigorous definition, or cannot be adequately articulated. This occurs when the processes are not well understood, or are still under development, or if the individuals involved lack the necessary communication and analytical skills.

Accelerated Pace of Work

Following integration into daily laboratory operations, the system gradually alters expectations for the timely completion of work from the laboratory. The elapsed time and amount of effort needed for many activities will be drastically decreased. Requests for information that may have previously taken days or hours to fulfil can now be done within minutes or seconds. For example, the elapsed time needed to ascertain the status of a sample and its testing is hastened with on-line information available from the LIMS database. The overall testing turnaround time also decreases through the automation of various information handling steps needed to process samples through the laboratory.

The improved timeliness of services generally stimulates increased demands. The technology enables existing staff to handle increased workloads. As a result the overall pace and intensity of work significantly increases.

Increased Capabilities for Group Self-management

The LIMS increases the measurability of work performed by various laboratory groups. It provides organized feedback on the quantity and timeliness of work performed by various laboratory groups. Details regarding samples and tests completed is readily available from the system's database. Prior to a LIMS, this level of feedback was obtained only through time-consuming and tedious clerical efforts. Since this information was generated manually, the timeliness and frequency of this feedback meant that each group obtained data that was weeks or even months old. Information provided by the LIMS provides each work group with the necessary operational data to manage the normal course of their own day to day activities. Supervisors and managers establish various procedural guidelines for each work group to act on information provided by the LIMS. Only instances that fall outside the established guidelines, the exceptions, are brought to the attention of supervisors. This provides each work group with greater autonomy over their daily routines and allows supervisors to concentrate their efforts on the management of exceptions and other conditions that affect the quality of services from the laboratory. Such a change facilitates the formation of what Peter Druker calls an 'information-based organization' in which there are fewer management layers and there is extremely rapid decision making.[2]

Changes in Job Skill Requirements

The laboratory will experience a gradual deterioration of skills formerly needed to carry out processes that have been automated by the LIMS. Laboratory personnel will have little reason to manually perform calculations or procedures that are handled by the new system. Their ability to complete these tasks will diminish as the need for these skills becomes less and less frequent. After a year or two, people will simply forget how work was done under the manual system.

---

[2] P. F. Druker, 'The New Realities', Harper and Row, New York, 1989.

Added Costs

The organization must bear the added cost for support and maintenance of the LIMS system. This includes items such as expendable supplies, maintenance agreements, and personnel responsible for daily operations of the system. A detailed presentation of post-implementation resource needs are discussed in Chapter 15.

Flexibility

Changes are generally easier and more quickly implemented in non-automated systems. For manual processes, agreement among the various individuals involved with the proposed alteration is all that is required to implement a change. With a computer, additional effort and more individuals are needed to install the approved changes in the system's software programs and database. Because of the extra effort, the LIMS may force a more detailed and thoughtful evaluation of the merits of proposed changes and how other functions may be affected.

The ease with which changes can be implemented on the LIMS depends on the degree of flexibility or control that is configured into the initial design of the system. A system designed for maximum flexibility may be overly complex and difficult to use. It does, however, rapidly accommodate changes. One that is optimized only for current operating needs may be simple and quick, but changes may be extremely difficult and time consuming to implement.

3 Common Misconceptions

A LIMS is often viewed as a panacea for all laboratory information management problems and bottlenecks. Its ability to solve existing problems is not inherent in the technology itself: it is a function of how the LIMS is implemented and deployed. Automation does not inherently eliminate errors or problems with the systems they replace. If implemented inappropriately, they can, in fact, exacerbate problems. This section discusses the common misconceptions regarding a LIMS.

Paperless Operations

A common misconception is that, with automation, information will be entered and displayed mostly on terminals or other computer devices, thereby reducing or totally eliminating the use of paper in daily operations. While computer terminals are well suited for the rapid retrieval and display of a limited amount of information, paper is an ergonomically superior medium for the presentation of large amounts of information. Information can be presented in printed reports with higher information densities and a wider range of formatting options.

The improved data access and enhanced report formatting from a LIMS actually stimulates the demand and use of printed reports. One estimate is that expenditure on paper and printing supplies increases by approximately 30% with the introduction of automation.[3]

[3] R. R. Stein, *Chemometrics Intelligent Lab. Syst.: Lab. Inf. Manage.*, 1991, **13**, 15.

Immediate Productivity Increases

Productivity increases occurs when the system is well integrated into the daily routine of laboratory operations. In fact, the near-term effect of a LIMS is a decrease in productivity as operations are transitioned from former system(s) and skills to the new system being developed. The speed at which productivity improvements became apparent depends on factors such as:

- how well the system meets operating needs of the organization;
- how rapidly the system is implemented;
- the quality of training provided to those that will use the system.

Inherent Reliability of Automated Systems

Computers are nothing more than extremely stupid, but unquestionably obedient, servants who repeatedly and very rapidly execute explicitly defined commands. The detailed steps to be executed are determined by the system's software programs. The reliability of software depends solely on how well its designers stipulate the exact procedures and conditions under which the system needs to operate. Frequently, in an attempt to accelerate implementation, the effort of defining the system's requirements and design is either abbreviated or eliminated. Errors and other deficiencies with the software are identified when the system is in use by the laboratory. Usually these errors prove to be nothing more than inconveniences that can be resolved by modifications of the system. Occasionally, the errors result in improper decisions that adversely reflect on the quality and credibility of the laboratory's services.

Aspects of the computer hardware that are highly mechanical or extremely complex are also prone to failure. Hardware components such as disc drives, tape drives, and networks are the components most prone to failure. The effects of these failures can be made transparent to users of the system if the appropriate procedures and resources are in place to compensate for these events when they occur.

Effort Required for Implementation

The amount of effort required to install and implement a LIMS is usually underestimated, especially in cases where a commercial LIMS software package is acquired. Significant resources are required to define many aspects of laboratory operations needed to configure the LIMS. The actual time needed to enter the configuration details into the LIMS is relatively small compared with the effort needed to understand the product and to define areas of laboratory operations that are generally undocumented. A discussion of implementation resources and the work required for implementation are respectively presented in Chapters 10 and 11.

CHAPTER 7

Determining Technologies that are Relevant or Irrelevant to a LIMS

There are numerous automation technologies that support LIMS. Sifting through the myriad of new innovations is confusing. This confusion is exacerbated by increasingly sophisticated marketing techniques utilized by vendors of the new technologies. A pragmatic discussion of automation technology life-cycles and how to establish their relevance is presented.

1 Automation Technology Life-cycles

Products based on automation technologies generally undergo the following stages in their lifetime.

State of the Art Products

Technology that is new or considered to be state of the art presents previously non-existent benefits in terms of absolute performance, functionality, or economy. The pace of innovation is staggering, especially in the area of hardware. Innovations in processor technology, semiconductive materials, and precision manufacturing techniques have produced innovations every 6 to 12 months. Improvements in hardware have significantly increased performance while lowering overall costs. Software improvements are also being introduced constantly, although not at the same frantic pace as hardware.

Products based on new and innovative technology need to first establish their commercial viability. Many innovations fail to achieve commercial success because they do not address customer needs, are not sufficiently robust for continuous operation, or they are not cost effective. Vendors introduce new products to commercialize new technology based on marketing studies and assumptions that are not always supported by the realities of the marketplace. The product itself may be viable, but its commercial acceptance is compromised by the vendor's failure to adequately make, sell, distribute, or support it.

As an adopter of newly introduced products, you assume certain risks.

- You will uncover product deficiencies and provide the vendor with valuable assistance in the debugging, refinement, and development of the technology for widespread use. You essentially serve as an adjunct to their product development efforts. The time and effort required may compromise your own efforts, if you are constrained by strict timing objectives and limited resources.

* If the product fails to realize commercial acceptance, you must assume responsibility for support and maintenance normally provided by the vendor. Alternatively, you can abandon use of the product entirely and replace it with other products or technologies.

These risks are worth it if the new product holds the promise of substantial benefits to your organization. As an early adopter, your experience and expertise in the newer technologies will precede those of other organizations. This can provide you with a substantial advantage, in terms of capabilities and efficiencies, over other organizations who take a more conservative approach towards the acceptance of new products and technologies.

Established Products

Products achieve acceptance in the marketplace after they have:

* established a base of satisfied customers to serve as a reference base; and
* created satisfactory financial incentives for the vendor to justify continuance of its investment in the products development, sales, manufacturing, and support.

Once a product has reached this stage, customer feedback gradually results in enhanced usability and performance improvements. Vendors pursue further development and refinement to broaden the product's appeal and marketability. They do so in response to customer needs but also to stay ahead of competitors.

Mature Products

Mature products have a wide base of customers with in-depth knowledge and expertise in its operation. They continue to serve customer needs and through a prolonged period of refinement, they achieve high levels of reliability. Many of the initial deficiencies identified when the product was newly introduced and established have been addressed.

Products become mature when the technologies upon which they are based is superseded by other more recent technologies. Gradually sales decline as customers pursue the newer products.

Discontinued Products

Products that are discontinued are no longer promoted or sold. Further development and refinement is also discontinued. Support and maintenance costs increase as parts and expertise become increasingly difficult to obtain. The customer base declines as they migrate to the newer products and technologies.

Reasons for a vendor to discontinue a product may include the following.

- They may be acquired or merged with another firm. Objectives of the new venture and management team may not support continuation of the product.
- Financial performance of the product, while respectable, is not adequate to satisfy business objectives.
- The product competes, directly or indirectly, with other similar products offered by the same vendor.

2 Establishing Relevance

The relevance of specific technologies and products involves several factors specific to your organization including its:

- risk tolerance;
- aggressive or conservative position on the use of technology;
- budgetary constraints.

Your organization's tolerance for risk can be gauged by the speed with which it has historically embraced new, unproven technologies and incorporated them into its business operations. Newly introduced technologies are inherently more challenging to implement. In exchange, the early adopters are rewarded by being among the first to realize the benefits of the new systems. This is important for organizations in highly competitive and information-intensive industries such as pharmaceuticals.

Those with a more conservative stance usually wait for established and mature products. They prefer to adopt technology after it has been proven and implemented elsewhere. The followers attempt to minimize their risks by leaving leading edge efforts to others. They capitalize on lessons learned by the leaders.

An examination of your organization's position on technology adoption is a useful starting point for determining its tolerance for new technology and products. What is relevant to one organization may be totally inappropriate to another. A thorough assessment of needs and goals for the LIMS is an essential starting point in establishing the relevance of various technologies. Each component of the system should support the goals of the organization. Situations in which there is no obvious and direct relationship between proposed technical solutions and business needs should be carefully scrutinized.

3 Technologies to be Considered

This section discusses various existing and emerging technologies that should be considered in a LIMS.

Database Management Systems

A Database Management System (DBMS) is a software program which controls how data is created, stored, managed, and protected. It provides the means through which data can be shared by numerous users and programs. A good DBMS is important for a LIMS in which several individuals and groups are involved with the creation, analysis, approval, and updating of information on the computer.

DBMS technology has replaced the inconsistent, inflexible, and redundant data storage mechanisms of previous computerized file processing systems. In a file processing system, data elements belong to a particular program designed to meet a targeted and isolated need. A given data element can exist in several locations and formats on the computer. Each repetitious occurrence may exist in different formats and may even have slightly different meanings. Changes made by one program or user may not be reflected in all other instances of the data element used by other programs and users.

A DBMS centralizes the storage of data on the computer. A given data element is stored in a single location. Users and programs need only be concerned about what the specific data element is called. Changes made by one user are automatically reflected in the information presented to other programs and users. The DBMS controls how it is retrieved, updated, and stored.

Relational Databases

Relational Database Management Systems (RDBMSs) are noted for their flexibility. Unlike a normal DBMS, an RDBMS is not constrained by the way that the information is initially set up on the computer. The RDBMS maintains the relationships between the various pieces of information and how they are physically organized. You can easily reorganize the data in a number of different ways. These are called views. This allows you to use information from the database in a number of different, previously unforeseen ways.

Most traditional non-relational DBMS are propriety vendor products. Each has its own unique set of instructions and commands needed to make it work. The newer RDBMS products, irrespective of vendor, utilize a standardized approach for working with information. This is called SQL, or the Structured Query Language. This makes it much easier to understand and work with several different RDBMS products.

The cost of the added flexibility of a RDBMS is somewhat lowered performance over its traditional DMBS and file processing system counterparts. However, adequate levels can be achieved through careful selection of the systems hardware and by skillful configuration of the database.

Multi-media Databases

Most DBMS and RDBMS systems deal primarily with numeric and textual data. However, laboratories are constantly overwhelmed by other types of data in the form of spectra, pictures, sounds, and electrical signals. Multi-media databases organize the various forms of data within the laboratory and provide a single and consistent mechanism for their access and management. Challenges of this technology include the relatively large size of each data element and their impact on the overall system and network performance.

Networking

Computer networks provide a way of sharing data and hardware resources. They allow for the distribution of work among various specialized systems scattered across several groups and locations. Networks are extremely important for an organization to realize the potential of increasingly powerful personal computers and workstations. Each computer performs specialized tasks for which its hardware and software has been optimized. Through the network, it can share information with other systems. This provides a tremendous increase in the organization's computing power and lessens its dependence on the larger, more expensive, minicomputer and mainframe systems. For a laboratory, networks facilitate the movement of data between instruments, the LIMS, and other corporate information systems.

Hardware

Hardware includes all the working and tangible elements of the computer. It faithfully performs instructions encapsulated in the myriad elements of the system's software. Each year, spectacular advances in hardware technology produce improved performance at substantially lowered costs. The cost of a computer system purchased in 1981 will, in 1993, buy a system with a 40-fold increase in processing power, a 1000-fold increase in storage capacity, high quality monitors, and other advanced features that were inconceivable in 1981! Anticipated revolutions in hardware design are expected to substantially increase performance and, at the same time, lower costs for systems introduced during the 1990s. The lower costs and higher performance greatly increases the affordability and accessibility of computers. This, in turn, promotes widespread use of automation technology by organizations and individuals who previously could not justify its investment.

Hardware performance improvements have facilitated the use and acceptance of increasingly sophisticated and easy to use software. Many of these programs were either too slow or were constrained by hardware limitations. There is generally a time lag of several months to a year between the introduction of new hardware technologies and the establishment of software capable of capitalizing on them. Relevance of the newer hardware to your needs should be considered relative to the availability of the software capable of meeting those needs.

User Interface

The type of people using computers has gradually shifted from programmers and the technically inclined to office workers, managers, and scientists who have neither the time nor the interest to learn programming languages or arcane keyboard commands. Systems that were intuitively easy to use have gained increased commercial acceptance. Systems based on a Graphical User Interface utilizing icons and a mouse, allow users to work through programs in an easily understood and intuitive manner. Technologies such as Virtual Reality and Data Visualization provide scientists with new ways of presenting, analysing, and drawing conclusions from extremely complex sets of data.

Systems Engineering

Systems engineering methodologies encapsulate comprehensive techniques for the implementation of automated systems. This involves a chronological sequence of distinct phases in developing an increasingly detailed understanding of the system's needs, its design, and the actual software implemented.

The methodologies have emerged to address needs to improve the efficiency, quality, and performance of automated systems. They are based on models that depict characteristics of the organization and the system to be automated. The models offer a combination of graphical and textual elements. Besides computer systems, they are also used to support total quality management and business process re-engineering initiatives.

Several methodologies are supported by automated software tools, called CASE for Computer Assisted Systems Engineering. The models are iteratively reviewed, corrected, and validated before the system is implemented. Model changes are easier, and less costly, to make than changes made to an installed system. Disadvantages are that the methodologies are constantly evolving, they take a long time to learn, and that there is a severe shortage of people proficient in their use.

In actual practice, the resultant models can be extremely rigorous. Practical methodology guides are published in workbooks of 10 to 20 volumes! Their creation can consume a lot of time and effort. Benefits include more efficient and higher quality implementations. The rigour and level of detail varies with the size and complexity of the LIMS. However, the benefits of these techniques are not realized if the people involved lack the basic training and experience in the methodology.

Structured analysis techniques were introduced in the late 1970s by Yourdon–Demarco, Gane and Sarson, Ward–Mellor, and Warnier–Orr.[1-5] Structured analysis is characterized by graphic models in the form of data flow diagrams. Each diagram is accompanied by textual process specifications and data dictionaries which amplify the detailed steps and information relevant to each process. The predominant emphasis of structured techniques is on data flow diagrams and processes.

Information engineering methodologies emerged in the early 1990s and have

[1] T. Demarco, 'Structured Analysis and System Specification', Yourdon Press, Englewood Cliffs, NJ, 1978.
[2] C. Gane and T. Sarson, 'Structured Systems Analysis: Tools and Techniques', IST, New York, 1977.
[3] E. Yourdon, 'Modern Structured Analysis', Yourdon Press, Englewood Cliffs, NJ, 1989.
[4] P. Ward and S. J. Mellor, 'Structured Systems Development for Real-time Systems', Yourdon Press, Englewood Cliffs, NJ, 1986.
[5] J. D. Warnier,' Logical Construction of Systems', Van Nostrand Reinhold, New York, 1981.

been popularized through works by Martin[6–8] and CASE software vendors such as Texas Instruments and KnowledgeWare. The analysis of data plays a dominant role in information engineering methodologies, and processes play a subordinate role. Information engineering is based on several fundamental premises.

- Most organizations manage their data in ways that are redundant, undefined, and inconsistent.
- An organization's processes are much more dynamic and changeable than its data.
- Processes are best supported and are the most adaptable to change if they are based on a consistent and well managed base of data.

It is imperative to realize that these methodologies represent tools and techniques, not stepwise recipes and dogma. They provide extremely useful ways for you to understand laboratory operations. The models provide a blueprint of the system to be automated. Their primary function is to facilitate communications between those charged with the implementation and those who will ultimately be affected by the LIMS. Secondly, they serve as reference points for those charged with post-implementation maintenance and support. As such, the models should be easy to understand, verifiable, and easy to correct. Your primary emphasis should be placed on obtaining a clear understanding of the laboratory, not on strict adherence to the gospel of any single methodology.

[6] J. Martin, 'Information Engineering I: Introduction', Prentice-Hall, Englewood Cliffs, NJ, 1990.
[7] J. Martin, 'Information Engineering II: Planning and Analysis', Prentice-Hall, Englewood Cliffs, NJ, 1990.
[8] J. Martin, 'Information Engineering III: Design and Construction', Prentice-Hall, Englewood Cliffs, NJ, 1990.

CHAPTER 8

Establishing Realistic Goals for the LIMS

Establishing realistic goals is a critical step in successful implementation of the LIMS. Goals that are established should enhance the laboratory's value to the enterprise and should be realized within a reasonable period of time.

1 The Importance of Goal Setting

Goals establish the overall business purpose to be fulfilled by the proposed system. They express management's expectations of benefits to the organization, independent of whatever technology is employed and how it is installed. Goals should be set by management of the laboratory and other groups within the organization who, directly or indirectly, are affected by laboratory information. They should be concise; a listing of one or two pages (or less) should be adequate. They should be expressed in terms understandable by someone with no prior knowledge of computer technology. Benefits targeted for the LIMS should be directly linked to the organization's Critical Success Factors, which are the key things that the overall organization needs to focus on to be successful in the face of demands imposed by customers, the competition, and regulatory bodies.

Clearly defined goals serve as the focal point of all subsequent aspects of the LIMS implementation. They are the principal input for the definition of LIMS requirements and costs (Chapter 9). As the project proceeds the cost, effort, and feasibility of meeting each goal are clarified. During the course of implementation, it is worthwhile periodically completing a management assessment of each goal in the light of newly discovered information regarding costs, effort required, and obstacles.

In situations where there are several goals, each should be assigned a priority relative to the others. This serves as a basis for the resolution of issues that will certainly arise in the future. Goals frequently conflict with one another. Assigning priorities helps in choosing between alternative, and sometimes equally viable, courses of action. As an example, two common goals of a LIMS are to improve laboratory efficiency and to shorten turnaround time for analytical testing. Generally these two goals are compatible; by improving the efficiency of selected operations, the net result is that it takes less time to complete the entire process. However, in some cases, the two goals conflict with one another. Improving service cycle times may require additional work and less efficient utilization of the LIMS, laboratory equipment, reagents, or other resources.

Factors impeding successful implementation of a LIMS or any other form of automation include failures in:

- goal identification and articulation;
- management sponsorship and guidance;
- involvement of laboratory staff;
- planning;
- allocation of appropriate levels of resources and expertise;
- competence in execution by the project team.

LIMS generally do not fail because of inadequacies in the technologies or the product offered. They more often do so because of how the implementation is directed, organized, and executed. Establishing clearly defined goals provides a basis on which to ensure that the various technological offerings are appropriately used to fulfil the overall business objectives of the organization. Goals ensure that the effort and expense associated with implementation is appropriately targeted so that results can be achieved in a timely and efficient manner.

Establishing goals for the LIMS is management's responsibility. Too frequently, managers abdicate to subordinates who are unaware of the organization's overall business imperatives and challenges. This is a big mistake! The result is systems that are never used, that do not address business needs, or which take years (even decades!) to implement. In the absence of clear guidance from management, efforts will be directed toward a variety of special interests and technical solutions which may not be compatible with improving overall operations of the organization.

2 Why Implement a LIMS?

A LIMS can provide both tangible and intangible benefits to your organization (Figure 8-1). Tangibles can be measured by quantifiable criteria such as direct reductions in current expenses, avoidance of expected increases, and analytical service cycle time improvements. They can be realized by reducing the effort and time for repetitive tasks and paperwork. Intangible benefits cannot be evaluated by objective criteria, but nonetheless, result in overall improvements to the organization. They include enhancements in how the overall enterprise and customers perceive the laboratory's competence, quality of services, and technological leadership. Intangibles often result from increased data consistency, reliability, and improvements in how information is reported.

In considering a LIMS, it is important to examine how your laboratories operate. LIMS technology generally applies to segments of a laboratory that have well defined procedures and testing protocols. There is a clear case for widespread use of a LIMS in highly protocol-driven environments. This includes certain aspects of manufacturing, product development, and some analytical service functions. The case becomes less clear in situations where protocols and methods do not exist or are still in the early stages of development. This is

---

A LIMS Can Benefit Both the Laboratory and the Enterprise

❏   Increased laboratory throughput

❏   Reduced testing turnaround time

❏   Labour savings

❏   Avoiding future staff increases

❏   Improved data management

❏   Improved laboratory management
  →   turnaround monitoring
  →   workload monitoring
  →   resource allocation

❏   Improved data accuracy

❏   Improved data consistency

❏   Improved format of analytical reports

❏   Time savings for laboratory clients

Figure 8-1 *LIMS benefits*

especially true for laboratories involved in basic and exploratory research. In these circumstances, technologies such as collaborative computing or electronic laboratory notebooks may be better suited. However, a LIMS may fit well into selected segments within these environments, as long as there is a minimum level of protocol definition and repeatability. This is especially true where there is a high volume of samples, data, and testing activity.

There are two reasons for a LIMS. The first is to overcome existing problem areas or constraints in the current way of doing things. This may be because of existing manual or automated systems that are ineffective, inefficient, obsolete, or that cannot cope with current demand levels. Has the organization experienced problems with regulatory compliance, data consistency, retrieval of data from archives, analytical turnaround, or sample status tracking? The second reason is to introduce new automation and management capabilities to the organization. Which aspects of laboratory operations and information management practices could be improved? How can the laboratory capitalize on capabilities afforded by automation technology?

General goals applicable to a LIMS are discussed in the following section.

They include goals affecting the overall organization as well as those specific to the laboratory itself. Goals relevant to your organization may not be included. Conversely, those presented may not apply to your environment. You need to carefully consider the exact goals that apply to your organization's needs and business imperatives.

Goals of the Overall Organization

The following discussion lists goals of the overall enterprise and presents ways in which the LIMS contributes to their realization.

Improved Product Quality

Analytical testing provides the basic information needed to confirm the quality of the organization's products and its processes. A LIMS can facilitate quality efforts by:

- improving capabilities for the orderly storage and expedient retrieval of test results and the supporting analytical data used as a basis for quality evaluations;
- reducing the cost of quality programs by improving the efficiency of analytical testing;
- providing an historical base of test data to monitor trends and variables associated with products and processes.

Customer Service

Analytical testing is often a part of the normal services provided to customers. It is also used in response to special customer requests or in complaint investigations. A LIMS can improve customer service by:

- reducing the amount of time needed to complete testing thereby reducing the overall customer service response time;
- providing an easily accessible historical base of analytical results regarding services provided or testing completed on behalf of each customer;
- reducing the time needed to transmit completed analytical results to customers by the utilization of electronic mail, automated facsimile transmission, or direct computer to computer linkages;
- decreasing or eliminating customer's efforts in the processing of paperwork for recurring laboratory support.

Product Improvement

Laboratories generate supporting test data for product improvements. This allows organizations to make their products lighter, safer, more efficient, and more durable. A LIMS can support these efforts by applying consistent management practices to the testing data and facilitating its use by those involved with product improvement.

Partnerships

Laboratories are increasingly involved with the maintenance of business partnerships between manufacturers and their raw material suppliers. In return for their patronage, manufacturers require that suppliers perform more testing and provide the results in an expedient manner. The amount of analytical work performed by the manufacturer is thereby decreased. They merely perform periodic audits and performance checks on their suppliers to verify the credibility of the analytical results. A LIMS can assist in these efforts by:

- establishing consistent formats for product certificates of analysis provided to customers;
- providing a basis for managing specialized certificate of analysis formats and customer-specific reporting requirements;
- providing an electronic means of transmitting analytical and other data to customers by mechanisms such as Electronic Data Interchange.

Efficiencies

Improvements realized by the introduction of a LIMS can contribute to overall efficiency improvements by:

- improving the utilization of manufacturing production capacity by reducing the cycle time for analytical testing upon which critical process decisions are made;
- decreasing the time and effort required to access, and make decisions based on, analytical information;
- establishing an historical and statistical basis to justify reducing the volume and frequency of routine testing programs.

Enhanced Data Analysis

The cost of analytical data is measured by the time, effort, facilities, equipment, and reagents required for their creation. The LIMS provides a basis for the increased utilization of this valuable resource through a variety of data analysis software tools. Included are capabilities such as chemometrics, location-based or geographic information systems (GIS), and process-based statistics such as statistical process control (SPC) and statistical quality control. The LIMS serves as a reliable, consistent, and accessible repository of analytical data which can be utilized by a wide variety of data analysis programs.

Laboratory Goals

The following discussion lists goals specific to the laboratory. Ways in which the LIMS contributes to their realization are discussed.

Productivity

The LIMS can improve the overall efficiency of laboratory operations by:

- reducing or eliminating the effort involved in specific information handling activities such as data transcriptions, calculations, and report preparation;
- expediting the dissemination of work assignments through the laboratory, thereby minimizing idle queue times for instruments and personnel;
- providing a basis to justify the acquisition of new resources or the shifting of existing resources by monitoring trends in personnel and equipment workload demands;
- improving laboratory resource management tools to dynamically monitor work input, work outstanding, work completed, and turnaround time performance of each laboratory group.

Quality

Improvements in the quality of laboratory operations and test data can be facilitated by the LIMS by:

- enforcing consistent operating procedures for such laboratory practices such as sample handling, data review, approvals, and reporting through the LIMS software;
- ensuring the consistent application of test methods in terms of the data recorded and calculations used;
- reducing recording errors by providing additional checks on the data as it is being entered into the system;
- improving the tractability of analytical data by providing intrinsic links to a responsible person, entry dates, and, if applicable, any reasons for data changes.

Improved Data Access Speed

Improved access to analytical data from a LIMS enhances operations of a laboratory by:

- facilitating the formation of self-directing work groups within the laboratory by providing easy access to information regarding workload demands and work group performance;
- improving the timeliness and effort required to respond to enquiries regarding the status of samples, tests, or other work within the laboratory;
- expediting the dissemination and use of analytical data by the laboratory as well as other groups within the enterprise.

3 Are Your LIMS Goals Realistic?

To be successful, the LIMS needs to realize its stated goals within a reasonable time frame. Implementations exceeding 2 or 3 years generally suffer from a lack of credibility, budget overruns, technical obsolescence, and defocused efforts. Meeting your goals requires a combination of automation technology and, more importantly, driving forces within your organization. Goals that are not supported by either the technology or the organization should be considered as incremental enhancements to the initial LIMS.

Determining if your goals are realistic can be done by answering two basic questions: will technology help and can the organization implement it?

The role of technology follows a translation of LIMS goals into an articulation of needs and a selection of system elements as discussed in Chapter 9. The relevance of technology is discussed in Chapter 7. An important part of this assessment is to review what similar organizations have done.

Your organization's ability to implement the technology and handle the changes required of a LIMS is fundamental to the realization of your goals. A small fraction (20–30%) is due to technology. The majority is a result of the organization's ability to focus the appropriate training, expertise, experience, funding, patience, and management direction to the effort.

CHAPTER 9

Needs Assessment and System Selection

The success or failure of your LIMS will be determined by how well it satisfies the information management needs of your laboratory. An up-front assessment of needs serves as the basis for choosing a system and how it will be implemented. The underlying premise of your implementation should emphasize solutions targeted towards the fulfilment of needs, not the installation of technology solutions awaiting needs.

The chosen LIMS will have a significant impact on how the laboratory will operate. It provides exciting opportunities for improving the efficiency and performance of the organization. The system will also make certain jobs more time consuming and difficult. A myriad of systems are available offering technological options. Items that should be considered in selecting a system include its breadth of functionality, expandability, development tools, performance, and other business considerations. Each option has several costs: the up-front purchase price, on-going maintenance costs, costs for services, and laboratory manpower costs. The laboratory's objectives, goals, and tolerance for risk need to be evaluated and compared with existing technologies.

1 The LIMS Needs Assessment

The LIMS needs assessment provides the supporting foundation for implementation of your system. The purpose of the needs definition is to establish *what* the LIMS is expected to do, without specifying *how* these goals are to be achieved. Information is gathered from the individuals and groups to be affected by the planned system. This information is assembled into a document to establish the organization's goals and quality criteria for the implemented LIMS. How well the delivered system meets these needs determines the success or failure of the LIMS.

Why is a Needs Assessment Necessary?

The needs assessment identifies multiple characteristics of the environment targeted for the planned LIMS. It encompasses factors that drive daily operations of the laboratory including items such as its business goals, long-term strategy, industry (or industries) serviced, work performed, analytical demands, regulatory requirements, operating hours, staffing, information handled, and logistical constraints. Implementations that focus on needs of the organization result in systems that enable revolutionary improvements in daily operations of the laboratory. Unfortunately, many implementations place too much emphasis on the installation of technological components of hardware and software, with little regard to how the technology adds value to the organization. These systems, while they may be technically elegant, are generally not well integrated with the people and the organizations for which they were intended.

The laboratory's inventory of needs provides a prospective basis for evaluating how potential solutions, in the form of software and other available technologies, can be applied to daily laboratory operations. It serves as a focal point for selection of a system and for planning the activities and resources for implementation.

Retrospectively, the assessment can define how well the implemented system satisfied the needs of the targeted organization. An inadequate definition of needs usually leads to improperly designed and poorly implemented systems.

Who Should be Involved?

A key element of a successful needs assessment is the inclusion of the appropriate people and groups from your organization. The people to be affected by your LIMS may present different, and often conflicting, views regarding what the system should do. In many cases, you will find disagreement and misunderstanding regarding current procedures and responsibilities. These issues are much easier, and more economical, for you to resolve before a particular hardware and software design is finalized.

Those that define LIMS needs should represent the various roles and responsibilities within your organization. Each individual may have one or more roles. This includes the following LIMS client categories whose needs must be satisfied.

- Operational clients
- Supervisory clients
- Management clients
- Executive clients
- Standard bearers
- Laboratory customers

Operational Clients

Operational clients will have the most frequent contact with the new LIMS. They are very much concerned with details regarding the individual functions to be performed and the manner in which they will interact with the new system. Details that are important to operational clients may involve issues that their supervisor may not be aware of, or interested in. However, meeting the needs of operational clients is critical to the success of the system. They may not have the authority to approve or reject decisions regarding the LIMS implementation, but the manner in which they use (or choose not to use) the new LIMS can make or break the system. It is extremely important that operational clients are directly involved in the needs assessment and all subsequent processes of the LIMS design and implementation.

Operational clients are knowledgeable about the specific job they do and the people with whom they have immediate contact. Their performance is evaluated by how well they do the specific tasks assigned to them. They may not be aware of how their job fits in with other activities within the group or of how their group fits in with the overall organization. In many cases, groups are segmented into several specialties and, within the same group, one set of operational users do not have a detailed awareness of functions performed by other operational users. It is important for you to identify and involve operational clients that represent the numerous laboratory functions to be impacted by the planned LIMS.

Supervisory Clients

Supervisory clients manage a group of operational clients and are responsible for their performance. Many formerly served in an operational capacity and are generally familiar with the work performed by operational clients. However, they may or may not empathize with the needs, concerns, and perspectives of their subordinates. This is because supervisory clients are evaluated by the performance of their unit as a whole, not by the narrower criteria that apply to operational clients. Supervisory clients are concerned with a number of separate functions and several different individuals interacting in a smooth and coherent manner. They are interested in the possibility of increasing the volume, improving efficiency, and reducing errors with the work performed by their unit.

Supervisory clients may or may not directly interact with the planned system. However, they are affected by it as long as functions and data handled by their group are processed by the LIMS.

Supervisory clients are directly responsible for hearing the concerns of operational clients. They are directly involved in conflicts with operational clients and serve as mediators of disagreements among other operational clients. They also serve as a conduit of communication from higher levels of the organization.

Management Clients

Management clients are responsible for the performance of several organizational units. They oversee the work of several supervisory or other management clients but generally do not directly supervise operational clients. Each organization may have several layers of middle-management. Management clients typically define timing and budget constraints for the LIMS.

Executive Clients

Executive clients are generally not directly involved in project details. They approve funding requests that originate from lower levels of management. Executive clients are typically more concerned with the strategic and long-term impact of the LIMS on the enterprise as a whole. This includes such items as the possibility of new markets, new products, new or improved services, or new capabilities to be gained from the system. They usually are not former operational or supervisory clients and typically are little concerned with the daily details of laboratory operations. Hence, they are not in a position to help to define needs for the people who will actually be using the system on a daily basis. However, they are able to define how the LIMS should fit into the overall capabilities and strategies of the organization as a whole.

Standard Bearers

Your organization may have people whose objective is to ensure that the LIMS is developed in accordance with various internal, industry, and regulatory standards. This includes people from departments such as the audit, quality assurance, validation, information systems, accounting, and regulatory affairs. These individuals are responsible for ensuring that the LIMS is compatible with the practices and technology applied to other areas of the organization. They represent the needs of the organization that are not represented by the operational, supervisory, management, and executive client groups directly affected by the planned LIMS.

Laboratory Customers

Laboratory customers utilize the services of the laboratory. This includes those that submit work and samples to the laboratory as well as those that utilize the results of testing.

Why Needs Should be Documented

Documentation of the needs forces clarity and discipline in communications. It facilitates the distribution and verification of this information by various client groups to be served by the LIMS. Having it in written form also facilitates the sharing of this information with individuals or groups that were not directly included in assembling the inventory of needs. The content of the requirements document needs to be understandable by a wide audience across a wide range of organizational and disciplinary boundaries. In the absence of proper needs documentation, this information is disseminated verbally, an error prone and inefficient process. Without adequate documentation of needs, it is easy for the designers, programmers, and others involved with the implementation to lose track of what the system was intended to do in the first place.

The Needs Inventory

The inventory of your organizations LIMS needs must account for the following.

● Functional needs
● Data needs
● Operating needs
● Maintenance needs

Functional Needs

Functional needs are represented by clear and precise statements describing what is to be done by the system. They do not specify details of the technology or programming languages through which functions will be implemented. It is important to segregate the underlying functional needs from the technical solutions that will be used to meet those needs. The exact determination of technical solutions is specified by an activity called design which evaluates possible solutions that encompasses the totality of needs.

Obtaining a precise definition of functional needs can be extremely difficult. Some people cannot segregate essential needs from the current (manual or automated) systems through which those needs are presently fulfilled. The physical manner in which business functions are executed are subject to revision with changes in technology, personnel, and facilities. The underlying business needs served by those functions generally remain essentially the same.

Data Needs

The data needs describe the exact information that must be handled by the system. This includes a listing of the types of information along with their respective definitions.

Operating Needs

Operating needs define the conditions under which the functional and data needs must be provided by the system on a daily basis. This includes specification of items such as sample volumes, hours of laboratory operation, number of staff to be serviced by the system, security needs, regulatory guidelines, and the need for independent audits. Also included are business operating needs such as the laboratory testing turnaround times and information storage retention times that must be supported by the system.

Maintenance Needs

Maintenance needs include a combination of functions, procedures, and personnel to ensure continued and reliable performance of the system following implementation in the laboratory. These include needs such as:

- procedures and personnel to provide timely assistance to those that use the system (your organization should specify the definition of timely response);
- resources to either fix or replace defective hardware and software components;
- documentation, which includes a description of the system design and instructions for its use;
- training programs for:
  - -- laboratory staff;
  - – in-house development staff;
  - – in-house system support staff;
- procedures, programs, and personnel to protect the data and programs residing on the system and to ensure total system recovery in the event of a system failure;
- procedures, programs, and personnel for arching historical information and the expedient retrieval of selected information from the archives;
- programs to monitor system performance and utilization of storage, communications, and processing resources.

The Needs Assessment Process

When the implementation is completed, those that use and are affected by your LIMS will, either formally or informally, perform an assessment of the system as they use it to support daily work activities. Systems that do not serve the needs of their targeted clients must be modified or they will eventually deteriorate and fall into disuse.

The speed and quality of implementations can be drastically improved if the project team has a good understanding of what the organization needs at the outset, rather than after the LIMS is installed and programs have been developed. Unfortunately, many LIMS teams do not complete an adequate needs assessment up-front. This activity is forgone due to the following common misconceptions.

Misconception 1: LIMSs are a standardized product.
Fact: The various commercial LIMS software products are each based on unique histories and targeted markets. Each product is based on assumptions regarding the working procedures and information handled by laboratories. These assumptions are based upon business practices employed by the laboratories upon which the software product's design was based.

Misconception 2: All laboratories have the same needs.
Fact: The needs of each laboratory is driven by the specific industry to which they belong as well as their role within the overall organization. The functions performed, information handled, and operating demands of each laboratory is different.

Misconception 3: Laboratory operations are simple.
Fact: Laboratories are often the most data-intensive operations in any organization. The quantity of data processed by laboratories typically dwarfs the information generated by all other parts of the enterprise. Much of this is due to rapid

advances in automated instrumentation and sampling devices that permit unattended continuous testing. Decisions that are extremely important to the organization are based on laboratory testing. This includes decisions regarding factors such as the adequacy of products and processes, safety of the environment, and critical life and death decisions on medical treatments.

A well thought out strategy for the needs assessment defines the manner in which information is gathered regarding the organization(s) targeted for the LIMS. Guidelines regarding this process are provided in the following paragraphs.

- Develop a plan for completing the needs inventory. The plan should include identification of the groups and, if possible, the exact individuals who will be involved. The contribution of each group and individual to completion of the needs inventory should also be identified.
- Obtain management approval of your plan. Advanced approval should be obtained from the managers of the people involved with your data gathering. Each organization has its own (usually unwritten) guidelines regarding restrictions on who you can meet with and when interviews should be scheduled. You can create an enormous political backlash by not obtaining some form of advanced approval.
- There are legitimate reasons for managers to review and approve your data gathering plan:
  - special arrangements may be necessary to ensure that the interviews do not interfere with the routine work assignments of the group.
  - your initial list of interviewees may include people who are unable to understand or articulate needs. Other individuals may be better suited as your sources of information.
- Conduct effective data gathering sessions. The quality of your needs assessment depends on the quality of the information gathered from the targeted LIMS client organization representatives. The quality of their participation will be directly affected by the manner in which you demonstrate your respect for their time and the value of their participation. The following guidelines are provided for effective data gathering sessions:
  - limit each session to 1 hour or less to ensure adequate focus and concentration.
  - provide at lease 1–2 days advanced notice of the topics, issues, and materials to be discussed.
  - gather and study any pertinent data prior to each session.
  - limit involvement only to individuals who can contribute or are interested in the topics to be discussed.
  - collate results of each session and verify the information gathered through either the distribution of minutes or follow-up meetings to ensure that you have correctly understood the information gathered. An effective technique is for you to state, in your own words, the outcome of the session and to have it verified by the participants.
  - avoid the use of technical jargon or acronyms. If they must be used, be sure that they are defined and thoroughly explained.

2 The Requirements Document

The requirements represent an attempt to document the various needs of the organization(s) affected by the planned LIMS. The documentation of needs is generally an imperfect and error-prone process. This is because of a shortage of individuals skilled in this process and the fact that many organizations do not recognize its importance.

It is normal for the needs document to undergo several iterations as it is reviewed, verified, and approved. Revisions to the needs document following approval is also normal in order to clarify previously undetected ambiguities, and to correct errors, or in response to changing demands.

The manner in which needs are documented may depend on the size of your planned system and the implementation standards adopted by your organization. In some situations, the needs documentation for a planned system is provided by several smaller documents, rather than in a single large one. The manner in which this information is packaged and documented is not as important as ensuring that the resultant work product clearly reflects the various needs of your organization.

The requirements are used to evaluate several possible solutions. They are used by vendors, information systems groups, or other parties to establish a technical and organization design regarding *how* the needs will be met. Total compliance with the stated requirements does not guarantee a quality system. If the requirements have not been properly defined, and errors exists in them, then compliance with requirements will produce a system that does not satisfy the organization's needs.

What is a Requirements Document?

The requirements document is an attempt to document needs to be met by the LIMS. The difficulty of this process cannot be underestimated because not all needs can be explicitly defined. Preparation of the requirement document is an attempt to maximize the quantity of explicit needs. Techniques for understanding laboratory operations as discussed in Chapter 3 are useful for documenting the organization's functional and data needs.

Purpose of a Requirements Document

The requirements document reflects the functions, performance, maintenance, and operations of the planned LIMS. It sets the basis for all subsequent phases of the systems design and implementation. Requirements serve as a starting point for the project team to evaluate and establish aspects of the system including the:

- hardware configuration;
- software configuration;
- networking and communications configurations;

- integration of the LIMS with manual procedures;
- expertise and effort required for implementation.

3 The System Selection Process

A LIMS is not a short-term investment. It will have a long-term effect on the operations and vitality of the laboratory for many years following installation. The selection process defines how information is gathered and decisions are made regarding the best solution to meet the organization's needs.
The objective of the evaluation process is to ensure that:

- the chosen system meets the needs of the laboratory and the enterprise. The obvious first step is to formally define the laboratory's needs in a requirements document as discussed previously in this chapter.
- the chosen vendor is qualified to complete the work. The best prices will provide little assurance if the chosen vendor is ill-suited to fulfil its obligations for installation and support of the system.

The best means of choosing the most qualified and cost-effective system is through competitive procurement. This generally occurs with the issuance of a Request for Proposal (RFP) to potential vendors from a qualified bidders list. The bidders then submit a proposal which describes how the bidder intends to meet the requirements and the cost(s) involved. The customer then evaluates each of the competitive bids according to predefined criteria and chooses the proposal which best meets the technical and business objectives of the organization.

To fulfil the intended purpose of free, open, and fair competition among vendors, the entire process must be carefully managed. Government agencies are generally obligated by statutes dictating how the procurements are executed and documented. Other organizations have various levels of policies and guidelines for this process. Before proceeding, you should determine the procedures and practices applicable to your organization. The essential tasks in this process are described in the following sections.

Establish List of Qualified Vendors

Create a list of candidate vendors that offer products or services that may meet your needs. The listing includes vendors known to have the technical acumen, financial soundness, and business expertise to do a good job.

The list can be formulated from trade shows, journals, and references from colleagues. However, several organizations have formal procedures to establish qualified bidders for all goods and services. In this event, vendors who have not been formally approved through these procedures are not allowed to receive an RFP. In this instance, you need to allow for additional time and effort to investigate prospective suppliers of the LIMS; especially in instances where your organization has no previous business relationship with one or more of the suppliers.

Establish Vendor Evaluation Criteria

Establish the objective and subjective criteria upon which the competing proposals will be evaluated and the final vendor selection is made. The selection criteria should be based on a combination of functional, technical, financial, timing, and other considerations important to your organization.

Prepare Request for Proposal

The RFP should be well structured and comprehensive. It should clearly define the requirements of the system. Sections of the requirements document can be included within the body of the RFP. In other instances, the requirements document can be attached as part of the RFP. The RFP also describes the logistical details of the procurement process, such as:

- where the proposals should be sent;
- the number of copies required;
- the deadline for submission of proposals;
- the organization and content of proposals;
- who to contact for additional details;
- a description of the proposal evaluation and vendor selection methodology;
- explicit statements regarding ethical practices and procedures expected from all vendors;
- the expected date for selection of a vendor.

The benefits of using a competitive process with a formal RFP are as follows.

- It indicates to the vendor community that you are genuinely interested in the procurement of a system. Only serious customers go through the time and expense of preparing a comprehensive RFP and requirements document. Vendors will give more attention to the preparation of proposals to prospects that demonstrate the diligence and competence to understand their needs before acquiring a system. Such customers have higher levels of success and can potentially be used as future references.
- It ensures that all vendors start with a common baseline of information and provides for fairness in the selection process.
- The competitive nature of the process ensures that vendors will be aggressive in pricing.
- It forces the vendor to come forth with definitive statements of products, services, and work products that will meet with the requirements set forth in the RFP.

Release RFP to Qualified Bidders

All vendors on the qualified list are provided with a copy of your RFP at the same time. This ensures that all are working with the same information and have an equal amount of time to provide a response. All vendors should be provided with advanced notice of the distribution date so that they can ensure that the necessary expertise and resources are available to review the contents of the RFP and create their proposal.

In some circumstances the RFP will contain confidential and proprietary information regarding the organization's operations, plans, and philosophies. Such information should not be disclosed to your competitors but it is necessary for vendors to provide a responsive bid. In these situations, protecting your organization's interest is best accomplished by requiring vendors to execute a confidentiality agreement prior to the receipt of the RFP. The primary purpose of this agreement is to restrict the bidders use and distribution of the RFP exclusively to the preparation of your proposal.

In a like manner, some of the information that you request from bidders may involve confidential aspects of their products, personnel, finance, and strategies. In some instances, vendors will refuse to disclose this information. Others may require you to complete a confidentiality agreement.

Respond to Bidder Enquiries Regarding RFP

Following receipt of the RFP, bidders may have needs for clarification or additional information. Comments, questions, and criticisms from vendors regarding the RFP and your requirements can greatly assist in strengthening the overall process. In some instances, your requirements and procedures may be poorly conceived or vaguely defined. You should establish a procedure for bidders to submit questions and a procedure for you to distribute answers on any technical, business, or logistical issues that may arise. Both the questions posed by bidders and your response should be distributed to everyone to ensure equal access to the same information.

Evaluate Bidder Proposals

A team representing your technical, management, and user communities participates in the evaluation of proposals. A common approach and set of criteria should be applied uniformly to the review of all proposals.

Do not base your choices purely on cost alone. The vendor's costs are one of many cost elements in the project. A low cost of hardware and software from a vendor can be offset by higher internal costs for implementation and support. The least expensive system may not meet the functional needs of the organization and may not be used, or it may not fulfil the goals of the planned LIMS. Costs should only be considered in differentiating systems of roughly equal capabilities.

Several approaches can be taken for proposal evaluation.

- Criteria of minimum functional and technical standards are established. The proposal that meets the minimum standards at the lowest overall cost is selected.
- A listing is made of all requirements and a relative weight is assigned to each. The degree that each proposal meets each requirement is assessed by the

assignment of a numerical score. The score and relative weights are multiplied and the vendor with the highest cumulative score is selected.
- Vendors are assigned a relative score for each decision category. The vendor with the best score is selected.

Pitfalls of the Selection Process

Common problems that inhibit completion of the selection process are:
- a poor understanding of needs. Little time and attention has been spent in arriving at an agreement on the needs to be served by the planned LIMS. Some of those needs that have been identified are described in too vague a manner.
- a lack of agreement on evaluation and selection criteria. Instead of discussing the merits of each proposal, a significant amount of time is spent determining the strategy to be used for selecting a system.
- improper communication of needs. The needs that have been identified are not documented and not distributed to bidders. The resultant proposals are incomplete.
- vendors basing their proposals on capabilities and technologies that are still under development which may or may not be ready in time to meet your timetable.
- expectations of complete objectivity. The selection process will always contain elements of subjective judgments. A great deal of time can be spent in over-analysis and attempting to discriminate between subtle differences in proposals.
- poor coordination and communication with bidders. The quality of the proposals from bidders depends on providing them with timely and accurate information.
- inconsistent bidder feedback. Unless a format is defined up-front, the contents and organization of bidder proposals will differ radically. This increases the difficulty and time needed to evaluate proposals.

Vendor Offerings

Vendors offer several products and services to meet the needs of your LIMS. Included in these offerings are hardware, software, and services. Rarely will all items needed for your implementation be provided by a single vendor.

Usually, vendors provide specialized products and services to meet their targeted markets. A general statement cannot be made regarding the exact combination of products and services offered. This has, and will very likely continue to, change rapidly. What vendors offer are driven by rapidly evolving, and notoriously fickle, market demands and trends. The last decade has witnessed a trend towards increasing specialization, with vendors concentrating their offerings on products and services that are consistent with their core expertise and their investor's desire for profitability. It is common for the various components of the LIMS to be obtained from several specialized vendors.

Expertise not obtained from vendors, for reasons of cost or availability, can optionally be obtained from individuals or groups within your organization.

Vendors may offer packages of products and services directed towards a specific niche, such as LIMS, for a given price. Because the items are related in some way, and the vendor has experience and expertise in the area, the package price offers a cost savings over separate purchase of individual components. This practice is called bundling. It offers price savings and convenience to the customer and a strong position in a specialized niche for the vendor. Elimination of specific components of the bundle, and the accompanying costs, are generally negotiated on an individual basis.

Elements of vendor offerings, including the following items which are likely to be included in vendor proposals, are discussed further in the sections that follow.

Software

The detailed instructions regarding functions of the system are provided through the software. This includes items such as the operating system, database manager, and numerous applications programs. Each year, the complexity and power of software increases substantially. The development of software is a resource-intensive process. Vendors have become increasingly specialized in their software product offerings. It is not unusual to have the operating system, database manager, and applications programs developed by different companies.

Hardware

Hardware includes the tangible items of the system such as terminals, printers, data storage devices, and the computer processor. All elements of hardware are usually available through the major computer vendors. However, computer hardware with specialized capabilities is available from specialized vendors.

Communications

The connectivity between different computer systems, which may be spread over several locations, is provided through a combination of networking hardware, software, and services. Various options are available depending on the amount of data, transmission speed, geographic and political separation between systems, and levels of reliability required from the connections. Numerous vendors are available to assist with the implementation of each option. This includes vendors of specialized networking hardware and software as well as those that provide the necessary transmission lines connecting the various locations.

Services

Services include personnel with the necessary experience and expertise to assist with specialized aspects of the installation and implementation of the system.

Vendors can supplement the resources and expertise available within your organization. Included are services such as:

- hardware configuration;
- software configuration;
- development;
- training;
- integration;
- networking;
- software support;
- hardware maintenance.

Hardware Installation. Upon delivery, the various hardware elements are unpacked and installed at designated locations within your facility. This requires detailed knowledge of proper installation procedures and functional testing to ensure that the hardware is working properly. This service is generally provided with the initial purchase of hardware. However, additional support may be required for installation of specialized hardware configurations or for installations outside normal working hours. Special assistance with hardware configuration may also be warranted if the hardware is being moved from one location to another.

Software Installation. The various software components are installed on your hardware. Various functional tests are performed to ensure that the software is working properly on your system. The cost of installation and initial software configuration is normally provided with the initial software purchase. However, since a typical system may have many software components, it is normal for several parties to be involved with the installation. The sequence in which each software component is installed is important. Certain software elements, such as the operating system, must be installed on the computer before other software can be installed. In the event of both the hardware and the software being purchased from a single vendor, software installation occurs at the vendor's facility and the hardware is delivered to your site with the software previously installed.

Software Configuration. Software programs have features for tailoring the way it works with needs specific to a particular system. Vendors normally provide a minimum amount of tailoring with each software purchase. This normally involves sufficient work to ensure that the installed software can demonstrate minimum functional acceptance testing criteria which serves as the basis for payment. A considerable amount of additional work is needed for the software to meet the needs of the organization. For a LIMS, this involves items such as the loading of your test methods and specifications for each laboratory. The majority of this work is normally completed by your LIMS project team. However, vendors can provide your staff with assistance and guidance in configuration of the software.

Development. Development involves the creation of functions that are not directly met by the purchased software packages. This includes the generation of new screens, reports, and databases. Vendors may provide program development services directly or may recommend other parties who can provide these services.

It may be necessary for you to purchase special software programs so that special functions can be developed on your system. Many programs are provided in a run-time version only. This means that the program will only support functions and databases that are an inherent part of the software package. The development of other functions and databases requires the additional purchase of development versions of these programs.

Training. Contemporary LIMS software packages are extremely sophisticated programs with powerful capabilities. Unfortunately, most purchasers fail to fully utilize the capabilities of their software due to a lack of understanding regarding its use. It is common for 80–90% of software capability to remain unused.

Specialized training courses are available directly from vendors. This includes training in the LIMS software as well as the other components of the system such as the operating system, database, development programs, and system management programs. Other vendors can develop specialized training programs to meet needs unique to your implementation. This includes services such as curriculum development, training materials, and specialized courses.

Integration. Systems integrators specialize in bringing together the products and services from several vendors to meet your needs. They specialize in blending myriad technologies commercially available along with the capabilities for the development of specialized programs that are not met by existing software. Systems integrators implement the necessary technologies for disparate hardware and software offerings from different vendors to work together in a coherent manner.

Networking. Networks allow different computers at different locations to work together. Many options are available depending on the distance between systems, the type of computers involved, and the desired communications protocol between systems. Specialists are available with the expertise to develop a configuration, install it, and maintain networks.

Software Support. Vendors of software packages provide services to assist customers in the use of their software. Each vendor defines what is offered in a standard service offering package. Services beyond those provided in the standard package are also available at additional cost.

Hardware Maintenance. Hardware vendors provide service packages for maintenance of their systems. These services are also available from several specialized vendors who provide maintenance for computers manufactured by the major 'hardware vendors.

4 Selecting the Approach

Having obtained responses from bidders and identified the available offerings in both products and services, there are several possible approaches for implementation of the LIMS. These include:

- use of a LIMS software package;
- in-house development;
- contracted development.

It is normal for combinations of each approach to be used in any given LIMS implementation. The approach taken generally evolves as the system is implemented. A common strategy is to start with a commercial LIMS software package followed by development, either through contracted or in-house resources, to provide for needs not served by the package. Others start with total in-house development of all LIMS functions and convert these selected portions of systems into commercial packages.

Use of Commercial LIMS Software

Commercial LIMS packages are supplied by the major analytical instrument manufacturers and scores of software firms. Instrument vendors have come to view laboratory information management as an integral part of their product offerings. Software firms see LIMS as an attractive niche market. The large number of available sources for LIMS software ensures that these packages are competitively priced.

Commercial products are often based on software developed and previously installed in a specific laboratory. These include screens, reports, and databases for a variety of general laboratory functions. The software is under constant evolution with periodic enhancements of functionality added to broaden the appeal and applicability of the product. All commercial packages are, by design, targeted to address general needs for a wide variety of laboratories. In doing so, the vendor needs to make assumptions about the data and functions carried out by laboratories. These assumptions may or may not apply to your laboratory.

Most packages are configurable. This allows you to tailor specific aspects of the vendor's package without developing software. Examples include the capability for you to define your test methods, expected results, calculations, specifications, security, program access, sampling schedules, clients, laboratory groups, and sample types. General-purpose screens and reports operate on the items defined specifically for your laboratory. Changes in the way your laboratory operates are implemented by utilizing the configuration capabilities of the LIMS software package, not through the development of programs.

Several packages utilize commercial database management software and are themselves created by commercial development packages. These LIMS software packages can be readily enhanced by the addition of new databases and programs using the same, or other, development capabilities.

Many LIMS software packages are based on proprietary databases and development tools. These databases and tools are utilized exclusively for the specific vendor's LIMS applications. The availability of development software and expertise for proprietary products are not as extensive as for the LIMS applications based on commercial databases.

In-house Development

This involves the actual coding of screens, reports, and other functions not provided by commercially available software. Software development is a complex and time-consuming process. Software is normally developed to augment functions or databases not provided by commercial software packages. This requires the training of in-house staff in the various development capabilities available for the LIMS. This involves an investment in:

- formal training courses, along with any related travel expenses;
- unproductive time while in-house staff acquire proficiency in the newly learned skills;
- additional software for program development.

However, some laboratories choose to develop all aspects of their LIMS needs. The principal disadvantage of this approach is the elapsed time needed to write the software and the large amount of effort required. Large software development efforts are notorious for cost overruns, delays, and failure to meet goals. The risk and level of effort involved is not justifiable unless the system can be implemented in approximately ten laboratories and the organization has a strong management commitment to provide the resources necessary to implement and maintain such a system. The principal advantage of such a system is that the laboratory has a system specifically designed to meet their needs.

Contracted Development

Software development services can also be provided through contractors. These companies are experienced in the development and implementation of software systems. However, only a few contractors specialize in LIMS.

CHAPTER 10

Resources for LIMS Implementation

Installation of a LIMS dramatically affects the way your laboratory will operate.[1] It may do so in ways that were not anticipated when you set forth on the implementation process.[2] Every person in the laboratory will be affected in some way. A successful implementation requires the orderly execution of both technical and procedural changes within your organization.

The technology is provided by components of the system's hardware, networking, and software. Its sophistication and magnitude depends on the specific configuration of your system. People with the appropriate skills in the various technologies are needed for implementation. These skills may already exist within your organization. Otherwise they can be obtained from external sources, or developed internally.

The specifics regarding how this technology can or should be applied to your particular laboratory require decisions that can only be made by your organization. This requires collective decision making of those intimately familiar with your laboratory. This involves things such as:

- do we keep the current system, or portions of the current system, after the LIMS is running?
- do we resolve inconsistencies in our current system or do we carry forth those inconsistencies with the LIMS?
- who will be the users of the system? Do we limit it to the laboratory or do we allow laboratory clients to access the system?
- where should we locate the terminals and printers? Do we have the space?
- which laboratory group do we implement first?
- how long do we keep results on-line with the LIMS?
- will changes be needed in the way that laboratory services are requested? Can submitter groups outside the laboratory accommodate these changes?
- do we standardize the way results are reported? If so, what should that standard be?

---

[1] R. D. McDowall, in 'Laboratory Information Management Systems: Concepts, Integration, and Implementation', ed. R. D. McDowall, Sigma Press, Wilmslow, Cheshire, 1987, p. 34.
[2] D. O. Bassett, *Am. Lab.*, 1987, **19:9**, 28.

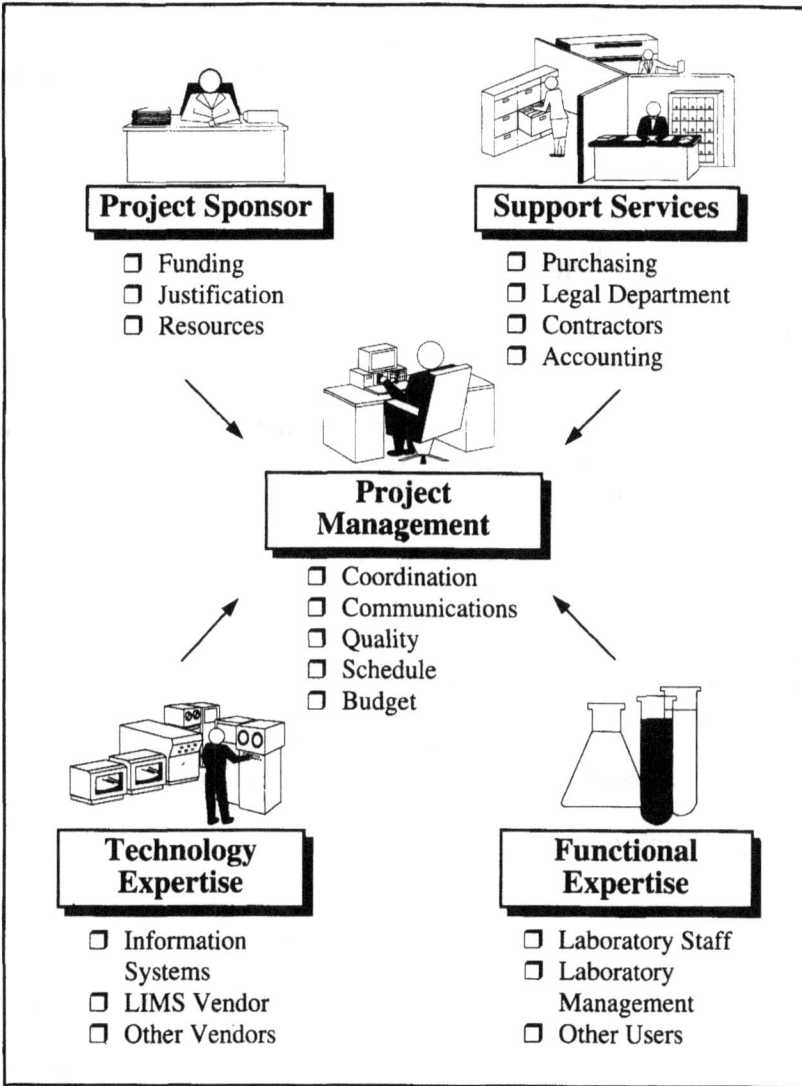

Project Sponsor
- Funding
- Justification
- Resources

Support Services
- Purchasing
- Legal Department
- Contractors
- Accounting

Project Management
- Coordination
- Communications
- Quality
- Schedule
- Budget

Technology Expertise
- Information Systems
- LIMS Vendor
- Other Vendors

Functional Expertise
- Laboratory Staff
- Laboratory Management
- Other Users

Figure 10-1 *Expertise required for LIMS implementation*

Implementation of the technological elements can usually be handled by a core group of highly motivated and appropriately trained people. Decisions regarding changes to the organization should not be left to the technologists. Representatives from your laboratory should be intimately involved with these considerations.

The LIMS team should be composed of at least two groups; one responsible for implementing the technology and another responsible for adapting the technology to laboratory operations (Figure 10-1).

Major responsibilities of the LIMS project team are described below. Each role may be fulfilled by one or more individuals. Conversely, a given individual may be required to fulfil more than one role. The number of people and the level of effort required varies substantially for each LIMS. The exact resources needed for a LIMS depends on the scope and complexity of the planned system. At a minimum, the project requires a management sponsor, a coordinator charged with maintaining project control and communications, and the direct participation of laboratory staff.

1 The Project Sponsor

The project sponsor is responsible for project justification, funding, defining goals, establishing timing objectives, and resolving project-related issues and conflicts. The approval and support of a project sponsor is required to satisfy the project's resource needs for equipment, people, and funding. He or she also ensures that sufficient participation is obtained from necessary individuals within and external to the laboratory.

The project sponsor, also referred to as 'the LIMS champion', provides the essential driving force, sense of urgency, and vision necessary for successful implementation. To be successful, he or she must have the proper authority and influence to ensure that the entire effort remains focused on meeting overall business goals and needs. This requires dedicating a small percentage of time to project reviews and setting overall direction.

Effective use of the LIMS often requires changes in the way the laboratory operates. The project sponsor critically assesses the value of such changes and provides the leadership and incentives to make the changes happen.

2 LIMS Project Manager

The project manager coordinates project resources, monitors progress of work efforts, tracks utilization of budgeted funding, and communicates with laboratory management and the project sponsor. The project manager is also responsible for keeping laboratory management and the project sponsor informed of progress and problem areas.

Primary responsibility for the work necessary for implementation rests with the project manager. He or she serves as the principal focus for communications and coordination of all groups involved in the effort. This requires a high level of skill in team building and leadership. Successfully completing project-related activities depends on the collective effort of many individuals, many of whom do not directly report to the project manager.

To remain on track, the LIMS should be the project manager's primary responsibility during implementation.

3 Technology Expertise

Hardware Vendor(s)

Hardware vendors normally provide for the installation and acceptance testing of newly purchased hardware. Vendors also provide annual maintenance agreements to support and repair hardware components that are already installed.

Software Vendor(s)

Software vendors provide services to assist with implementation of their packages. This includes the LIMS as well as other software applicable to your needs. The exact services provided by each vendor changes in response to technical and customer needs. A few of the services typically provided are:

- installation of the applications and other supporting software directly acquired from the vendor;
- formal training course(s) on the applications software;
- technical support for the applications software;
- maintenance agreements and periodic updates and enhancements of the applications software;
- additional training, consulting, development, or implementation support services.

Information Systems Group

The contributions of a central Information Systems (IS) group varies significantly from one LIMS to another. Their participation largely depends on the particular role assigned to IS within your organization.

Ways in which the IS group can participate in the LIMS effort include, but are not limited to:

- installation and support of network services across several sites;
- coordinating hardware installation;
- providing systems management services;
- assuming responsibility for hardware maintenance;
- providing resources for software development;
- providing technical reviews of the project;
- establishing corporate-wide development and integration strategies;
- providing project management leadership for the implementation.

Development

If necessary, development specialists install features beyond functions provided by software packages. This includes software development activities such as system design, data modelling, database definition, and the actual coding of programs.

Systems Operations

A systems manager coordinates all activities necessary to ensure reliable operations and availability of the computer. This includes coordination of internal or vendor resources for performing activities such as hardware maintenance, system performance monitoring, configuration management, system security, database maintenance, backups, system recovery, and the execution of scheduled preventative maintenance.

Validation

Validation specialists define the applicable standards and regulations for the LIMS. This generally depends on the industry served and the systems criticality to laboratory operations. Validation specialists ensure that appropriate measures have been taken for validation testing, site installation, qualification, documentation revision control, and independent audits of conformance with standard operating procedures.

Training

Training specialists establish training courses, perform user training sessions, and prepare user manuals. They ensure that new users are appropriately trained and qualified to complete designated functions on the LIMS.

Support

Support specialists provide direct operating assistance to users of the system. This involves detailed operating instructions in the use of the system and its programs. It also entails answering enquiries on the capabilities and operations of the LIMS. In addition to end-user support, they identify and catalogue areas that impede the usability of the system.

4 Functional Expertise

Laboratory Management

Laboratory management ensure that appropriate staff are available to assist with the implementation effort, they participate in the issue identification and resolution process, they ensure that project goals are realized within their own organization, and they notify the project manager and project sponsor of factors impeding implementation. Their support and involvement, or lack of it, critically affects the outcome of the LIMS.

Clients of the System

Designated user liaisons serve as the principal interface between the LIMS implementation team and the ultimate clients of the system. This includes groups both within and external to the laboratory. They impart their functional knowledge of operations into the implementation process to ensure that the technology remains focused on business needs. User liaisons also serve as the focal point of communications within their respective areas. They are chosen because of their detailed knowledge of both the workings, personnel, and culture of groups who will be affected by the LIMS. In some cases, user liaisons also coordinate and participate in the testing and validation of the system.

5 Support Services

Purchasing Agents

An organization's internal purchasing department is normally responsible for formalizing purchase agreements with vendors. This includes issuing purchase orders, price negotiations, establishing contracts, and ensuring compliance with appropriate transactional regulations.

Engineering/Site Services

Your local engineering group may assist you in making any needed facility modifications to support the computer hardware. They may also provide assistance with the installation of computer cables linking the system to terminals and printers in the laboratories.

Accounting

Your accounting or financial services group can provide valuable assistance in preparing the financial justification analysis for your LIMS. This includes analysis of, for example, the system's return on investment, leasing *versus* buying, cash flow, and net present value. The involvement of your organization's accounting department will add credibility to the financial case for the LIMS, since they will apply the appropriate accounting standards and criteria acceptable to your organization.

Contractors and Consultants

Contractors and consultants can provide you with assistance in meeting temporary needs for support or provide expertise that is not available within your organization. This includes support in the form of technology expertise, project management support, advisory services, or a wide variety of miscellaneous needs necessary to carry the project forward.

The cost of using external consultants and contractors is generally higher on a per-hour basis than full-time employees. The cost differential decreases when the indirect costs of full-time employment are considered such as insurance, retirement, vacation, other benefits, and administrative expenses. Also, unlike employees, their services are easily terminated as soon as the work is done.

The exact costs of contractors and consultants vary considerably and are usually driven by market forces. Variables includes the exact area of expertise, skill level, experience, and the volume of work involved.

Costs can be controlled through well thought out contracts that specifically define the exact nature of services and tangible workproducts desired. The principle advantages of using contractors and consultants are that they are used only for as long as you need them, and that they may do the work more quickly because of their specialized skills and experience.

CHAPTER 11

Essential Elements of a LIMS Implementation Plan

This chapter presents a general discussion of project planning concepts as they relate to a LIMS implementation. The plan establishes the work, resources needed, and timing objectives. It provides metrics for monitoring progress and accomplishments once implementation has started. Details specific to a LIMS are also discussed including the systems life-cycle approach as well as strategies for integration of the system into the daily routine of laboratory operations.

1 Reasons for a Project Plan

A project plan provides the mechanisms for determining when the LIMS will be installed, how much work it will take, and how much it will cost. It is a way of estimating the length of time and effort needed for completion. Once it has started, the plan is used to monitor the project's resource utilization, progress, and budget.

A good project plan has other benefits. It can be used for you to efficiently keep track of work that needs to be done and to coordinate support from the various parties that ultimately contribute to the success of your project. A good plan allows you to simultaneously track several activities at once. It serves as an inventory of completed work, work still in progress, and work that needs to be done in the immediate future. If task completion is unexpectedly impeded, the plan provides you with a list of other areas that can be worked on to keep forward momentum on the project. It also provides a mechanism for determining how schedule changes in one area will impact other work.

The project plan defines the various steps needed to ensure that the organization's investment in the LIMS is well directed. It provides the organization with a means of monitoring progress and monitoring the rate at which funds are consumed. It serves as a tool for identifying problem areas as or before they develop, so that corrective actions can be initiated.

The project plan is a tool for estimating, coordinating, reporting, and communicating. It gives you a basis to support answers to questions such as:

• when will the LIMS be installed?
• how much will it cost?

- how much have we spent so far?
- how much time will you need from me during the next six weeks?
- when will my laboratory group be in LIMS training?

2 Elements of a Project Plan

The project plan consists of the following elements:

- *the work breakdown* which is a detailed listing of work that must be completed;
- *a workplan* which specifies the relationships between the listing of work covered by the work breakdown;
- *a schedule* which includes targeted dates for the initiation and completion of all work elements;
- *a budget* which includes how much and when funds are required;
- *a project team organization* which identifies the individuals or groups involved with implementation and their respective responsibilities.

Development of your project plan is an iterative process of refinement. In the early stages, your knowledge about the final outcome of the project, in terms of timing and costs, is not well developed. Initially, it is based on incomplete and imprecise data. As the project proceeds, your knowledge is refined as more information is gathered and you gain increasing awareness of both the capabilities and constraints of the various technical, organizational, political, and cultural elements of the LIMS and the project team. As your base of knowledge and experience increases, so does the accuracy your estimates.

It is imperative that your project plan reflects the detail that you, your management, and your organization need to feel comfortable that the implementation is adequately managed and controlled. You should not make the project planning and the project management process more complex and time consuming than it has to be. The plan's content and level of detail varies considerably from one person and one organization to another. Managers are generally interested in key milestones and when they are reached. Clients of the system need to know when they will be directly affected by the system. Those intensely involved with implementation tasks need even higher levels of detail.

A wide range of project management software is commercially available. It varies widely in capabilities, cost, and ease of use. It automates many of the time consuming routine aspects of tracking work status, resource utilization, recalculating schedules, and forecasting budgets. The software is also useful for generating various charts and other forms of analysis for reporting on the project's status. However, to become proficient in its use, it requires an investment of time and training.

A clear definition of the LIMS objectives and scope serves as a firm starting point for your plan development process. This will establish the project's overall goals and provide metrics for creating initial estimates of its duration, cost, and effort required.

- The *objectives* state the business needs to be met by the LIMS. Several common reasons for implementing a LIMS are discussed in Chapter 8. The LIMS objectives should be stated in a way that means their outcomes can be either quantitatively or qualitatively determined. Upon completion, performance of the system relative to these objectives determine if it is a success.
- The *scope* of the LIMS defines the functions, size, and number of groups to be served by the system. Also included are factors such as work volumes, types of testing, number of users, and testing programs to be included.

Each project should have a definitive beginning and end. The end is achieved when clearly defined and demonstrable goals are realized. In the absence of unambiguous end-points, it will be impossible for you to manage the costs, effort, and schedule. Just because the implementation has reached its designated end-point does not mean that work on the LIMS has ended. The LIMS itself will grow and evolve long after its initial goals have been realized. This is a natural consequence of the organization's and laboratory's desire to further capitalize on opportunities enabled by automation. It also occurs in response to new business needs.

Your organization may have adopted planning methodologies that can be used as a starting point for your LIMS project plan. You can use them or create your own based on the concepts presented in the remaining sections of this chapter.

The Work Breakdown

Numerous segments of work are required to implement your LIMS from the initial starting point to the targeted outcome. A good project plan segments the work into discrete, manageable pieces. Tangible work products against which progress can be measured are defined for each segment. Each workproduct represents the outcome of efforts by one or more individuals working on a specific task or activity. Work products are usually in the form of documentation or reports including the facts uncovered, knowledge gained, or milestones accomplished.

A work breakdown represents one strategy for segmenting the total effort into smaller, easier to manage, pieces (Figure 11-1). Each major segment produces a tangible work product as an outcome. The segment is not complete until its accompanying work product has been completed and accepted. The total implementation is decomposed into several subprojects and numerous work products. A preliminary estimate of effort (hours or days) required is then assigned to each task.

The initial versions of your project plan will be based on incomplete information and many unknown factors. It is important that you identify areas in which more data is needed and that you define specific tasks to obtain this information. The accuracy of your plan is subsequently refined as this information is obtained.

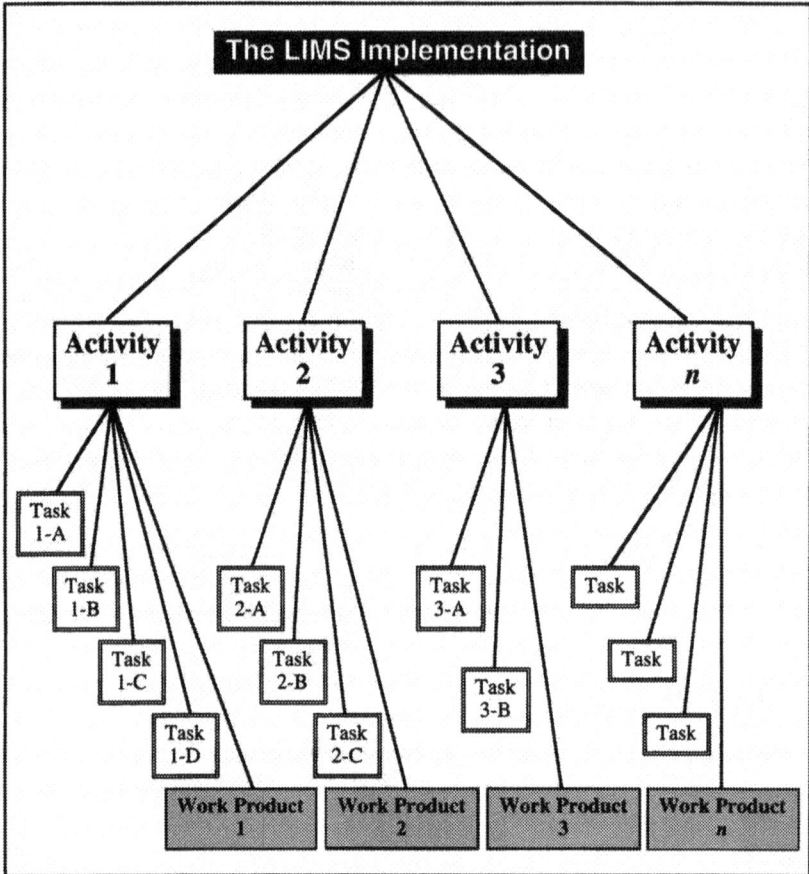

Figure 11-1 *A LIMS implementation work breakdown structure*

The Workplan

The workplan specifies the chronology and relationships of all work elements included in the work breakdown. To create a workplan you need to identify any work that must be completed before each task can be started. For example, the task 'install terminals' cannot be started until the task 'install network cables' has been completed. In establishing these relationships, you will probably find several items of work that you overlooked! A task network diagram (Figure 11-2) graphically depicts the chronological dependencies of the various work elements. Such a diagram is often called a Pert chart.

The Schedule

Once the specific elements of work are identified (the work breakdown) and their chronological relationships are established (the workplan), a schedule is obtained.

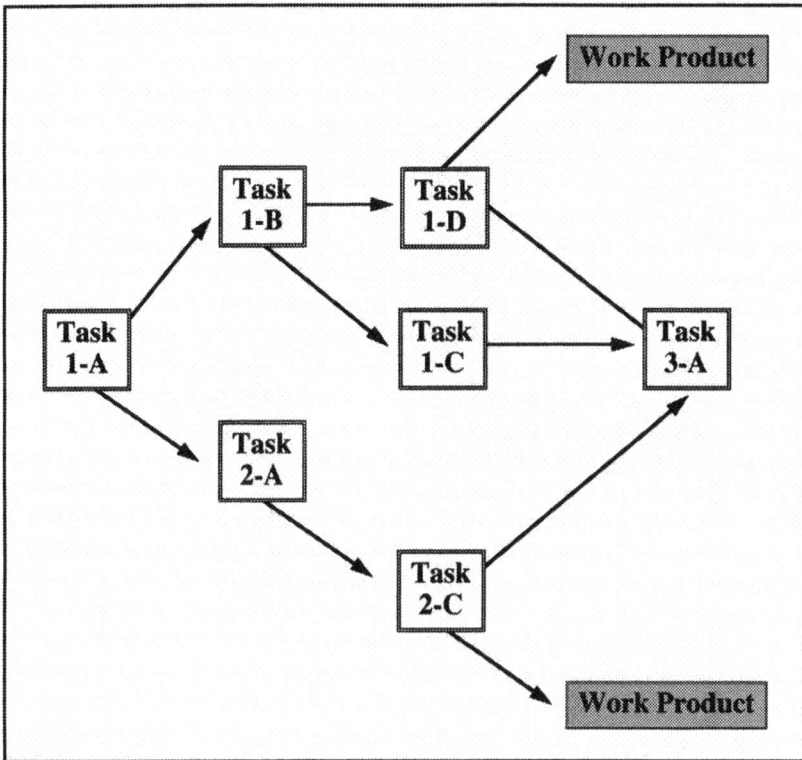

Figure 11-2 *Chronological task relationships*

This involves the following.

• start dates and end dates that are assigned to each task;
• specific individuals or groups are then given the responsibility for completing each task.

A Gantt chart (Figure 11-3) is one way of graphically depicting your plan to complete specific implementation activities. These charts can be created with the assistance of graph paper, specialized forms, or a variety of project management software packages.

Your initial schedule can be validated by assessing the amount of work assigned to each individual. This process, called resource levelling, involves balancing the amount of effort required with the availability of resources. Initially you will find several people scheduled for work that is beyond the total available work hours. At this stage it is not uncommon to find that certain individuals are scheduled for more than 24 hours of work per day! During other periods, the same individual or groups may be idle or underutilized.

Resource levelling ensures that the anticipated work demands are consistent with each individual's allotted commitment to the project. It involves changing task start dates, obtaining additional resources, or shifting the project's timing

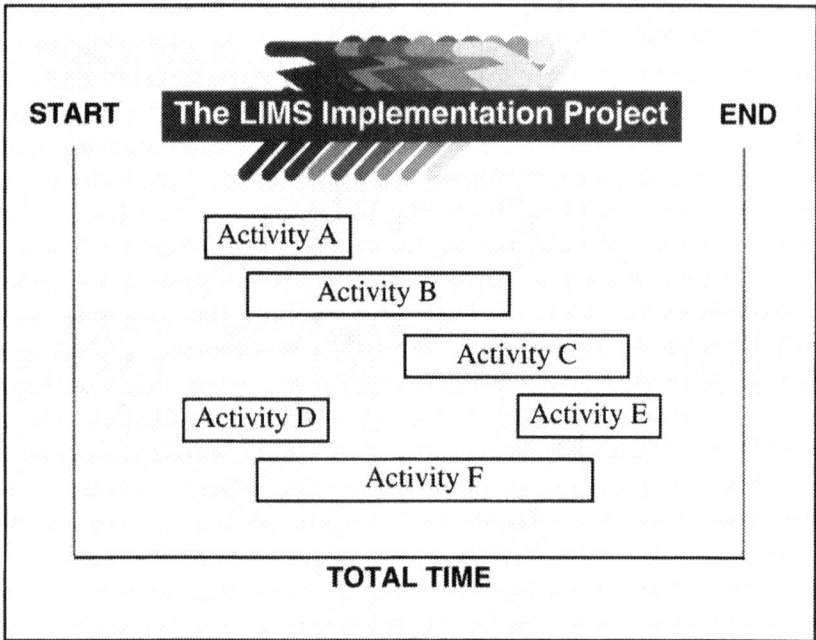

Figure 11-3 *Break-up of the project into smaller pieces*

objectives. Following several revisions, the outcome is a project schedule that specifies the targeted completion date for the overall project as well as the proposed start and end dates of each underlying work element.

The Budget

The budget includes all funding required for external purchases from vendors. This includes items such as the computer hardware, software, cables, training courses, manuals, *etc*. The budget also encompasses the costs of personnel required for the effort. This includes not only the hourly costs of their salary but also the overhead costs associated with each person's benefits and administrative expenses. The indirect overhead costs of labour generally range from 40 to 60% of the salary.

The project's budget includes a breakdown of various cost elements and a schedule for the consumption of funds.

The Project Team Organization

Initially, your proposed project team organization defines, in very general terms, the types of expertise and resources needed (Figure 11-4). A LIMS usually requires contributions from many groups and individuals within and outside your organization. The groups that need to be involved are first identified. Next, individuals from each group are identified.

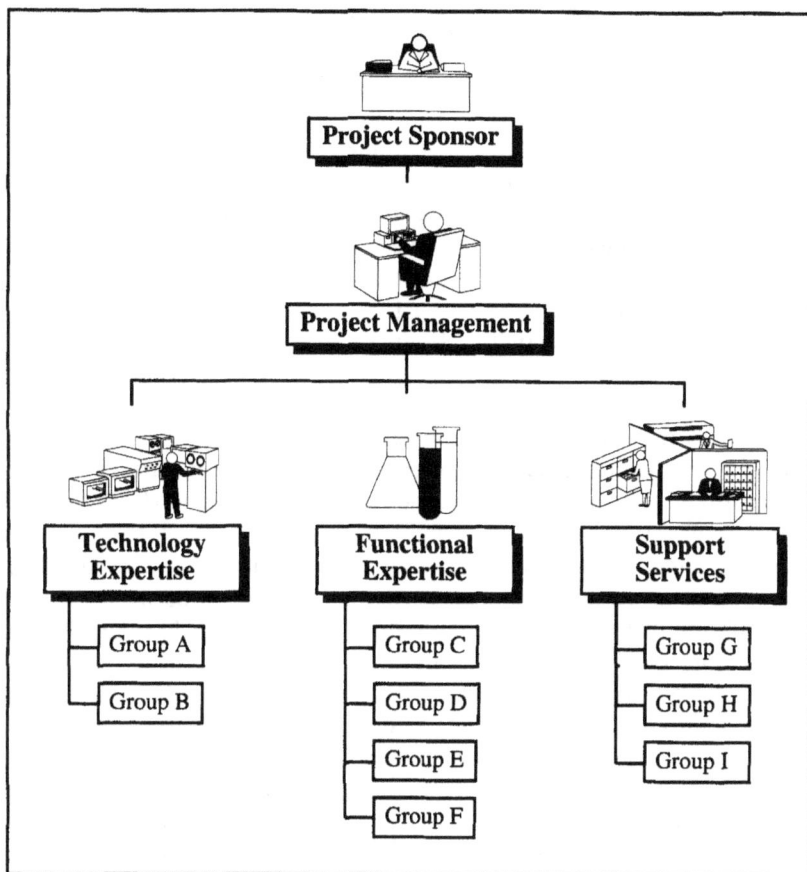

Figure 11-4 *Overview of a typical LIMS project organization*

The vast majority of people involved will contribute to the LIMS effort on a part-time basis, in addition to other job responsibilities. The exact levels of total effort required depend on the size and timing goals for your implementation. A LIMS that is large in scope or that has ambitious timing objectives requires the full time attention of several individuals.

The objectives and scope of your LIMS system is a good starting point for identifying parties who should play a role in implementing the system. If it is targeted for five laboratory groups, then some way of coordinating with each group must be included in your project plan. If new hardware is to be purchased, then it may be necessary to involve your company's purchasing agents and legal department to complete the necessary transactions and contract negotiations. If existing hardware will be used, then the appropriate computer group must be contacted to ensure that it can handle the demands of the planned LIMS. Typical responsibilities of the various groups that may contribute to implementation are discussed further in Chapter 10.

3 The Systems Life-cycle Approach

The Systems Development Life-Cycle (SDLC) approach ensures that the system evolves and develops in an orderly and disciplined way (Figure 11-5). SDLC has been initially developed as a way of minimizing the seemingly uncontrollable costs, unreliable delivery schedules, and poor quality levels historically associated with software implementations. The overall approach involves changing the way systems are implemented. Instead of being crafted by a select and isolated group of computer scientists, the systems are engineered in a logical, chronological sequence. This approach emphasizes a focus on business requirements to be served by the system and not on the elegant or clever details of the hardware and software technologies.

The philosophy of the life-cycle approach is based on the initial creation of system models that implementers and users can mutually review, verify, and iteratively refine before actual work on coding the software starts. The models take the form of written text as well a wide variety of diagrams (Chapter 3). It is hoped that most errors will be detected and that revisions will be made to the models, not functions that have already been developed and installed. The models serve as mechanisms for resolving disparities between what the implementation team perceives as laboratory needs *versus* what the laboratory actually requires to fulfil their daily business functions.

An analogy can be made between construction of a building and implementation of a computer system. Both start with a definition of needs to be fulfilled. The building may serve as a residence, a commercial establishment, a warehouse, or an office. A conceptual model of the building evolves in the form of diagrams such as blueprints and other documents. They are extensively reviewed with the targeted occupants to ensure that the design and construction details are consistent with their needs. The conceptual model and detailed design of the building are formalized before work begins on the site. Changes to the blueprints are much easier and less expensive to make than tearing down a wall and starting over. A typical LIMS project similarly proceeds through phases where the system is modelled before actual work on the software and hardware begins. The underlying premise of this approach is that changes to a model are much easier and less costly to implement than changes to a system (or building) that has already been constructed.

However, a computer system is markedly different from a building. A building is composed of an assembly of tangible objects. Progress can be readily evaluated by the appearance of physical structures such as a foundation, a basement, walls, floors, and a roof. Completion is marked by something that can be seen and touched. For a LIMS, assessing progress and completion is not as straightforward. The only tangible aspect of every computer system is its hardware. However, solely completing installation of the hardware, such as terminals and printers, does not necessarily mean that the LIMS is complete. Other elements still need to be in place before the LIMS system can accomplish useful work in the laboratory.

The generic implementation phases described previously provide a sequence

System Development Life-Cycle (SDLC) Phases

Phase	Results
PLANNING	Feasibility Assessment Initial Estimates
NEEDS ANALYSIS	LIMS Goals Scope Definition Constraints Needs Inventory Revised Plan (40%)
DESIGN	Hardware & Software Selection Automated Lab. Operations Design Technical Design Implementation Protocols Cost Justification Revised Plan (15%)
CONSTRUCTION	Configured Software Developed Programs & Databases
INSTALLATION	Validated Hardware & Software Training
MAINTENANCE	Continued Reliable Operations

Figure 11-5 *The systems life-cycle approach*

for incrementally refining your concept of what the LIMS will be. At each phase the models are reviewed with appropriate parties from the laboratories and other affected groups. Revisions to each model incorporate additional information and further levels of detail. The models are ultimately used to establish the systems hardware and software. Each phase has one or more work products to serve as input into the subsequent phase of implementation.

There is no standardized way of implementing the life-cycle approach inherent in the SDLC. The names and ordering of these phases are different in numerous development methodologies and in standard approaches endorsed by various organizations. The breakdown and activities listed in this section merely provide critical concepts that should be considered as you develop your own implementation strategy.

The flow of implementation does not follow a predefined sequence as implied in Figure 11-2. In fact, activities from different phases may overlap or run simultaneously. During the course of the project, errors and new information will surface that may require corrections to knowledge gained from previous phases. The phases provided here are merely groupings of activities with a similar purpose.

This section provides a generic breakdown and accompanying descriptions of work needed to complete implementation of a LIMS. The various activities are grouped into distinct phases listed below. This breakdown is consistent with a traditional implementation life-cycle approach for installation of software systems.

Planning

The planning phase establishes the feasibility of the LIMS. It sets forth the initial goals, scope, and strategy for your LIMS. Its purpose is to obtain a rough approximation of what the LIMS will do for the organization as well as order-of-magnitude estimates of resources needed for implementation. Your organization will review these preliminary estimates to determine if the LIMS is worthy of further investigation. Information obtained from planning activities will be refined in the needs analysis phase.

Typical planning activities are included in the following sections.

Survey LIMS Installations

Complete a review of systems implemented in laboratory environments that are similar to those within your own organization. You can identify them through a variety of sources: through published case histories, meetings, or professional contacts within your industry. Assess the workings of each laboratory and how closely it relates to operations of your own organization. In each case, determine their targeted goals, technologies utilized, how much it costs, how long it took, and benefits realized. You should ask the question, 'in retrospect, what would you do differently?'. The answers should provide valuable insight regarding your planned LIMS.

Survey LIMS Technologies

Survey the contributing hardware, software, and networking technologies of LIMS products. Chapter 2 describes the various components of a LIMS. Chapter 7 discusses the various technology elements underlying a LIMS. The purpose of this activity is to gain a cursory familiarity with the terms, concepts, and product offerings supporting LIMS.

Establish LIMS Feasibility

Gather and assess information within the organization to determine if further work on a LIMS is warranted. Develop a preliminary estimate of its goals, benefits, and costs. Create a preliminary project plan as an initial estimate of implementation resources, timing, and strategy. If the analysis proves promising, then the subsequent phases of 'needs analysis' and 'design' will refine estimates of the projects deliverables, cost, and timing.

Needs Analysis

The needs analysis phase serves to define, in non-technical terms, what the LIMS will do for the organization. The results of the analysis phase serve as the basis for formulating technical hardware and software solutions which occur during the design phase. During the needs analysis phase, it is important to remain focused on the needs of the organization and not to get too involved with the details of how they will be implemented on the hardware and software. The outcome of the needs analysis provides further refinement on the project's benefits, costs, and timing. If done properly, estimates obtained from this stage should be plus or minus 40%.

Typical activities included in the needs analysis phase are included in the following sections.

Establish LIMS Goals

The overall goals to be targeted by the LIMS are determined. These goals should be consistent with factors and strategies that advance the overall organization. A discussion of LIMS goals is presented in Chapter 8.

Define Scope of the LIMS

The following items determine the magnitude of the project:

- the laboratory groups to be served;
- the groups outside the laboratory;
- the groups outside the enterprise (*e.g.* customers, suppliers, regulators);
- the functions provided to each group, including data handling and reporting;
- the number of people within each group that will use the system;

- the various physical locations to be connected;
- the volume and type of samples handled by each group;
- the volume and type of tests completed by each group;
- the quantity of test methods to be converted;
- the types of samples;
- routine testing programs;
- routine sampling protocols;
- communications with other computer systems.

Identify LIMS Project Constraints

Management may dictate non-negotiable conditions on the LIMS. Other standard-bearers within your organization may also limit your options regarding how the system will be implemented. This typically includes groups such as information systems, legal, or regulatory affairs. Factors constraining a project include items such as its:

- timing goals;
- budget limits;
- technical standards for hardware, software, databases, and networks;
- software development, installation, and testing standards;
- project management standards;
- validation requirements.

Obtain Understanding of Current Operations

Establish the information and sample handling practices of the organization by obtaining a detailed understanding of its processes, data, and timing relationships. A good understanding of current practices and needs is important prior to determining where automation should be applied. Chapter 3 presents techniques for doing this. Chapters 4 and 5 provide a general discussion of laboratory operations.

Inventory Needs to be Fulfilled by the LIMS

Once you understand current operations, the next step is to determine the needs that will be fulfilled by the LIMS and those that will not. The result is a LIMS needs assessment and a requirements document, as discussed in Chapter 9. The management and staff of the organizations to be affected by the LIMS should participate in this process. A neutral facilitator or strong management representative should be involved in the unenviable chore of segregating needs of the overall organization from the desires of selected factions, groups, or individuals.

Refine Project Plan

Refine the project's estimated budget, resources, and timing estimates based on the new information (goals, scope, constraints, and needs inventory) gathered from the needs analysis phase. At this point, your cost estimates should be accurate to within 40%.

Design

The design phase details how the hardware and software technology will be applied to meet the LIMS needs specified in the analysis phase. It also establishes how operations of the laboratory will be affected by the new system.

Typical activities for the design phase are included in the following sections.

Select Software and Hardware

Based on the needs inventory and other products of the analysis phase, you should complete selection of the technical elements of the system. This includes the various elements of its hardware, software, and networking. Chapter 9 discusses details of the systems selection process.

Design Automated Laboratory Operations

The design defines how laboratory functions will be implemented with the selected software and hardware. To do so, you need to refer to the requirements and determine which laboratory functions will be implemented either through the selected LIMS software package, other selected software packages, or developed software.

The outcome of this effort is a detailed understanding of how the selected technologies will blend with the overall workings of the laboratory.

In addition, you need to establish how the various elements of laboratory data will be implemented on the system. This includes specifying, for each data element, those that can either be mapped directly to the LIMS software database, or mapped directly to databases provided with other applications, or those that require the development of special databases.

The design of automated laboratory operations has two aspects; one involving tangible items which intimately affect day to day operations of the laboratory, and another, less obvious one, that deals primarily with how the underlying technical elements are assembled. The first involves visible interactions that users have with the LIMS. Included are items such as the menus, screens, reports, functions provided, and how users navigate through the system. The second aspect is not visible to most users. It consists of the detailed programming elements, logic, and structures buried behind what is readily apparent to users.

The design process further defines the system in terms of both its affect on business operations and the technical elements necessary to implement the LIMS. It serves the needs of two audiences: the laboratory who must use the LIMS and the technical specialists who need to implement it.

Representatives of the laboratory must be involved with the design of the visible products that will be used in routine laboratory operations. The system will change the way that the laboratory does its work and services its clients.

Issues regarding the feasibility of any proposed changes and alternative courses of action will certainly surface as the design proceeds. The affected organizations should be involved in the discussion and resolution of these issues since new procedures and resources may be required.

Routine users need not be involved with the underlying technical details of the systems architecture, languages, programming structures, and code. The vendors, programmers, maintainers, and other specialists on the project team should address these issues. Confusion often arises when users are asked to participate in overly technical discussions on topics for which they have neither the training nor the interest. Users are mostly concerned about how the system will affect their work. Others involved with the LIMS implementation are responsible for the details of making the technical elements work to satisfy the organization's needs.

Complete System Integration Design

The system integration design determines how the LIMS interacts with other systems in the enterprise. It details how data is transferred or shared between systems. It also establishes how processes on one system activate processes on other computers.

Establish Implementation Protocols

Implementation protocols specify the detailed steps to ensure that all components underlying the LIMS are installed and used according to prescribed levels of quality and reliability. The exact protocols required depend on size, criticality, and the regulations applicable to your system.

The implementation protocols that may be required include those detailing the procedures, documentation, and responsibilities for:

- verification testing and validation of the LIMS;
- training of the project team and users;
- development, review, and approval of programs;
- control and monitoring of hardware and software revisions;
- maintenance of the system (Chapter 14);
- identification, review, and implementation of changes to the design (Chapter 14);
- back-ups and recovery of the system;
- independent quality assurance audits of the system.

Refine Project Plan

Products of the design effort provide the necessary metrics for estimating the capital, resources, and amount of effort involved with the remaining phases of installation. Using this additional detail, you should update the plan's organization, staffing, budget, and schedule. At this stage your estimates should be accurate to within 10%.

Complete Cost Justification

Complete a cost justification analysis for the LIMS as discussed in Chapter 12.

Construction

Construction of the system involves building functions to be used by the laboratory, either through the use of software packages or through the development of specialized functions.

Conversion

The conversion process implements existing laboratory procedures and operations on the LIMS or other applications' database software packages. Information specific to the targeted laboratory or organization must be defined and loaded into structures specific to each software package. This includes items such as the organization's products, customers, charge codes, sampling programs, test methods, and specifications.

The LIMS software package cannot be used by the laboratory until this information has been defined and loaded. To do so, you need to catalogue all documentation describing the laboratory's protocols, test methods, specifications, and other operations planned for the LIMS. You also need to define those items that are not formally documented. This requires a good understanding of the workings of the LIMS and other software packages on the system.

In most circumstances, the selected software packages do not completely fit operations of the laboratory. These instances should have been identified during the design phase. In these situations, implementation requires slight to substantial changes either to the way the laboratory operates or it requires the development of specialized programs.

Development

The development process creates functions, either through tools provided with software packages or through the creation of specialized programs and databases. The magnitude of this effort varies considerably.

The process should be guided by consistency in the tools utilized, documentation generated, standards employed, and control procedures. Users should be involved with the review and acceptance of all developed functions.

Installation

The LIMS installation is complete when it has reliably demonstrated its ability to be used for the routine processing of samples and testing within the laboratory. For this to occur, the hardware and software elements need to be installed. This also means that training has been completed for all users and that the system has been integrated into the day to day routine of operations.

Hardware Installation

The system's hardware must satisfy both the business objectives defined in the needs analysis phase and the technical details of the design.

The hardware should be installed by vendors or other qualified personnel. Ensure that the system passes the appropriate tests and diagnostics. If necessary, create the necessary documentation certifying appropriate installation of the hardware.

In some situations, the hardware may require a special room isolated from the solvent vapours and dust in the laboratory. Other situations may also require the installation of environmental control and power management systems. This is *especially* important if the laboratory is in a facility with extremes of temperature and poor electrical power quality. Hardware vendors generally provide environmental specifications for their systems. You should obtain a copy and take the necessary steps to ensure that your conditions comply with the vendor's physical hardware requirements. This includes factors such as:

- space needed;
- temperature;
- humidity;
- dust control;
- solvent vapour isolation;
- a sufficient quantity and suitable location of electrical power and outlets;
- quality of electrical power (voltage spikes, voltage levels, frequency);
- fire suppression systems;
- cables.

Software Installation

Software includes all the elements of purchased packages as well as any developed programs and functions.

Functions from developed programs outside standard applications software should be installed only after they have successfully completed functional testing in their pre-production environment. This may consist of a separate computer system, a physically separate installation of the database on the same computer, or special codes segregating the production from the pre-production data. This protects the working programs and data that are used by your laboratory on a daily basis. Furthermore, an acceptance test should be run after each software component is installed on the production system.

Training

A thoughtfully designed and executed training program ensures that people within the organization have the necessary knowledge and skills to reliably use the system. The system is essentially useless if those who need to use the software cannot apply it to their daily working activities.

Training first involves the development of a training program. This defines who will be trained, the specific subject matters covered, and the specific education techniques utilized; such as lectures, on-line exercises, manuals, reference cards, or self-paced instruction. The training plan also identifies possible sources of training. Vendors provide courses to support their own products. Topics relating to the specifics of your laboratory and its operations will need to be developed by the project team.

Maintenance

The appropriate maintenance program ensures continued reliable operation of the system. Chapter 15 discusses the responsibilities and resources for maintenance of the system following implementation.

4 Strategies for Introduction into the Laboratory

As a LIMS implementer, your job involves considerably more than selecting, purchasing, and installing hardware and software. To be successful, you need to fit the assorted hardware and software elements into the mainstream of laboratory operations.

Your plan for introducing the LIMS must be sensitive to the characteristics and needs of the affected laboratories and organizations. The needs are generally to maintain past levels of laboratory services while trying to integrate the new system into the organization. Your strategy for introduction of the LIMS into the organization needs to consider specifics such as:

- ensuring that the people and groups contributing to the LIMS effort are able to provide the needed levels of support;
- ensuring that extra time is allotted for either collective or individual review by all necessary parties;
- if required, providing the necessary time and resources for the creation, review, production, and maintenance of written procedures and evidence that they are being followed;
- if required, ensuring that time and resources are available for independent audits of the implementation process;
- providing extra time and attention to the training of groups with low levels of experience with automated systems;
- keeping various groups informed on progress and problem areas;
- ensuring that mechanisms are in place to capture, review, and resolve issues and concerns raised during implementation;
- establishing means of communication and coordination with project teams and laboratories which are geographically distributed.

Reasons for Partitioning the Project

You can realize incremental achievements by breaking up your LIMS implementation into smaller more manageable units. The laboratory and the management are probably anxious to realize some benefit from the LIMS as soon as possible. Partitioning your implementation project, as shown in Figure 11-3, provides one way of doing this.

The advantages of this approach are:

- that the laboratory realizes limited benefits early on, with more comprehensive solutions to follow later;
- that lessons learned during the earlier stages are applied to expedite the subsequent efforts;
- that it is easier to manage and coordinate several smaller and separate activities than one very large effort; and
- that the project team gains credibility and support from successful completion of the earlier activities. This facilitates work on, and funding for, the later efforts.

Strategies for Partitioning the Project

Your project can be partitioned in several ways. A few common approaches are discussed in this section. There is no single right way to partition your LIMS. The best solution for your circumstances is probably a combination of the various approaches described below.

Partition by Benefits

Benefits to the organization determine how the project is segmented. Portions with the highest potential benefits are done first. The others are done at a later time.

Partition by Ease of Implementation

Functions which are the easiest to implement through the software are done first. The more sophisticated, more automated, features requiring more effort are added incrementally.

Partition by Sample Type

The criticality or work volume associated with each sample type determines its sequence for implementation. Priority may be given to samples that are the most (or the least) critical. Alternatively, samples with the highest (or lowest) work volume may be given first priority.

Partition by Laboratory Group

Implementation sequentially proceeds one laboratory group at a time. The first groups are those with the highest (or lowest) staffing levels. This controls the number of people that need to learn the new system at the same time.

Partition by Costs

The LIMS starts with the smallest possible hardware and software configurations. It is gradually upgraded as more users and functions are added.

Implementation Options

Introducing the LIMS is complicated by the fact that the laboratory is required to continue processing samples and completing tests during the switch to the new system. An appropriate period of training and indoctrination is needed for the laboratory to become familiar with using it in their daily work. Laboratory services to the enterprise generally cannot be interrupted during introduction of the LIMS.

Murkitt[1] describes three modes of implementation.

Total Immersion

On a given date, the former system is discontinued and all new work is done on the LIMS. This requires complete confidence in the system gained through a comprehensive process of testing, validation, and training. This approach requires the total support of the laboratory's management and staff. It is best applied when there are substantial differences between the old and new systems.

Parallel Operation

The old and new systems are run simultaneously. This places a high work burden on laboratory staff. Results from the old system confirm performance of the new LIMS. During this process, it is not unusual to find discrepancies due to previously unknown errors in older practices! Complete transition to the LIMS occurs when the laboratory is confident of its reliability.

Selected Use

The LIMS initially automates a subset of laboratory functions. A small group within the laboratory demonstrate how the system works in automating daily laboratory operations. Other functions and groups are then added in carefully staged increments.

[1] G. S. Murkitt, in 'Laboratory Information Management Systems – Concepts, Integration, and Implementation', ed. R. D. McDowall, Sigma Press, Wilmslow, Cheshire, 1987, p.67.

CHAPTER 12

Justification and Approvals

The LIMS justification states the reasons why the system should be installed. It consists of a combination of both tangible (cost reduction or cost avoidance) and intangible benefits to the enterprise. The facts presented in the LIMS justification are used to obtain funding approval and authorization to proceed with the project. The LIMS cannot be realized without the necessary funding and without the blessing of the organization(s) to be affected by the proposed system.

This chapter presents considerations for justification of a LIMS and the process in which approvals are obtained. Details of the justification and approval process vary considerably from one organization to another. The roles of various groups that may impact the LIMS justification and approval process are also discussed.

1 Justification

The justification details the system's costs and the expected benefits that the LIMS will deliver to the organization. Data required for justification generally consists of:

- a comprehensive inventory of all LIMS cost elements;
- an analysis of current laboratory operating costs;
- an estimation of tangible and intangible benefits to current operations, both to the laboratory and to the various groups served by the laboratory;
- a comparative assessment of costs *versus* benefits.

LIMS Costs

The overall cost of a LIMS includes:

- initial purchase of the hardware and software;
- effort and materials needed to implement the system;
- support and supplies for continued maintenance of the system following implementation.

All too frequently, the costs of the LIMS are underestimated. In the past, many justifications have been based solely on the cost of the initial hardware and software purchases. Over the lifetime of the system, this represents less than half the total cost of the LIMS.

Initial LIMS Purchase Costs

Cost Category	Examples
Hardware	**Storage Devices**
	Terminals
	Workstations
	Printers
	Network Interfaces
	Bar Code Readers
Software	**Operating System**
	Database Management System
	Core LIMS Applications
	Optional LIMS Modules
	Program Development
	Instrument Interfacing

Figure 12-1 *Initial LIMS hardware and software costs*

System Purchase Costs

The cost of initial acquisition of the LIMS is the total sum of funds for the purchase of hardware, software, services, and other equipment. An example of initial LIMS hardware and software cost elements are shown in Figure 12-1. The exact hardware and software items appropriate for the LIMS are based on requirements established from the needs assessment (Chapter 9). The actual cost of these items are finalized with your chosen vendor(s) during the system selection process.

LIMS Implementation Costs

Implementation costs include the effort and materials necessary to establish the system in a working state that can be used by laboratory staff (Figure 12-2). Upon delivery, the LIMS consists of an assortment of hardware and software

Initial LIMS Implementation Costs

Cost Category	Examples
Internal Effort	LIMS Project Manager
	LIMS Project Team
	Laboratory Staff
	Support Groups
Purchased Services	Vendor Support
	Contractors
	Training
Other Costs	Travel
	Books
	Manuals
	Office Supplies

Figure 12-2 *Initial LIMS implementation costs*

tools that need to be moulded into a system suitable for assisting with daily operations of the laboratory. Resources are required to appropriately configure the system to meet needs specific to the targeted laboratory. This includes time required from laboratory staff as well as any assistance purchased from vendors. Furthermore, training is necessary for those charged with LIMS implementation as well as those that will eventually use the system. The cost of training includes the time for laboratory staff to participate in the training, the actual cost of each training course, and any travel expenses.

LIMS Operating Costs

Once implemented, the LIMS requires resources in the form of personnel, equipment, and supplies to maintain continued operations. These recurring expenses are needed to ensure that the system continues to operate in a reliable and consistent manner. The post-implementation resources and support cost elements are detailed in Chapter 15.

Individual(s) should be appointed to be responsible for management of the LIMS. Their duties include both preventative and remedial maintenance of the system. Preventative maintenance includes routinely scheduled items such as backups and archiving. Remedial maintenance involves restoration of the system and its database from unexpected failures in electrical power, networks, software, or hardware. Vendor-supplied hardware and software maintenance agreements complement, but do not replace, on-site support.

Support should also be provided for users of the system. This involves providing instruction on the use of the system and answering questions from novice users as they gain familiarity with the LIMS. Initially, end-user support needs will be high. Demands should diminish as laboratory staff gain familarity with the system.

Laboratory Operating Costs

The costs of laboratory operations without a LIMS is determined. This includes the amount spent for salaries, supplies, travel, telephone, postage, and other expenses. These costs are based on current procedures and the time required to perform laboratory tasks. Some of this data can be obtained from historical records. These costs are used as a baseline for the determination of LIMS benefits. The LIMS justification details aspects of current operating costs that will be affected by the system.

A detailed breakdown of time spent in specific laboratory tasks can be important in understanding the various dimensions of current labour costs. This information can be obtained either through special studies such as surveys or through time and motion studies. The time associated with each task is then multiplied by the staffs' salaries and overheads.

The most detailed analysis of a LIMS's effect on laboratory operations is presented by Cibulas.[1] in a Calgon Water Management case study. In a one month time study of laboratory activities, she found that 21% of the laboratory's effort was spent on information handling tasks that could potentially benefit from a LIMS. The anticipated benefit from a LIMS was estimated for each surveyed information-handling activity. The overall savings were estimated as 950 hours per month (Figure 12-3). The tangible justification of the LIMS was based on these projections. Other planned benefits included analytical turnaround time reductions, improved results reporting, and management control reports.

Tangible Benefits

Tangible benefits can be quantified and assigned a monetary value. They are realized from LIMS capabilities that either reduce expenses or which provide the

[1] A. E. Cibulas, 'LIMS: A Tool to Meet the Challenges of the 80s', Eastern Analytical Symposia, New York, 1990.

Projected Effect of LIMS on Monthly Laboratory Information Handling Activities

Activity	Hours	Projected Benefit	
CLERICAL			
Sample Tracking	44	50%	22
Time Allocation	80	20%	16
Work Order Preparation	128	80%	102
Filing	169	80%	135
Test Audit	23	80%	18
ADMINISTRATIVE			
Scheduling	26	80%	21
Administrative Reports	23	100%	23
Validate Reports	116	25%	29
Validate Time Sheets	16	25%	4
DATA MANIPULATION			
Transcription	227	100%	227
Computation	248	50%	124
Interpretation	253	50%	177
	1,560		953

SOURCE:
A.E. Cibulas, 'LIMS: A Tool to Meet the Challenge of the 80's'
Eastern Analytical Symposium, 1990.

Figure 12-3 *Projected LIMS benefits*

possibility of increasing sources of revenue. Each tangible benefit can be associated with a financial value that will be realized during a specified time frame. Tangible benefits are categorized as follows: cost reduction, cost avoidance, and revenue enhancement.

Cost Reduction

Cost reductions are realized by savings in current expenditure on labour, materials, services, or facilities. Included are LIMS functions that reduce costs by requiring less effort. Examples include savings on activities such as transcription, calculations, reporting, and statistical analyses. Net savings can also be realized by accelerating the elapsed time needed for the completion of targeted work activities. This decreases dead time and increases the utilization of equipment and personnel. For example, an improvement in testing turnaround time for a manufacturing operation can, in some instances, improve the utilization of facilities and equipment worth millions. Following implementation, cost reductions from the LIMS should decrease or eliminate specific elements of historical operating expenses.

Cost Avoidance

Cost avoidance involves a reduction, deferral, or total elimination of expense increases anticipated for predicted workload growth or the introduction of new capabilities and services. Cost avoidance benefits affect anticipated future expense increases unlike cost reduction benefits which are based on savings on current expense levels. The credibility of cost avoidance benefit assumptions are based on the certainty of occurrence of events leading to the projected increase. For example, a justification based on a laboratory handling a 20% volume increase with existing staff is suspect if historical data show consistent growth levels of only 10%. Hence a part of the case for benefits that avoid cost increases involves establishing a credible case that expenses will, in fact, increase without a LIMS.

Revenue Enhancement

Revenue enhancement include benefits that stimulate increased demands for the products and services of the organization. This includes features such as improved report formats, enhanced data quality, electronic data interchange, and more rapid response to customer enquiries. Increased business revenues occur if the LIMS provides features that distinguish the services and products of the organization as superior to those available from others.

Intangible Benefits

Intangible benefits are those that cannot be quantified due to a lack of hard evidence to determine their monetary value due to insufficient time and resources to quantitatively assess their impact. Examples include such features as better sample tracking, improved data organization, expanded access, enhanced quality of operations, extension of potential service capabilities, and improved laboratory management. A LIMS benefit that is tangible for one laboratory or one organization may be intangible for another. The only difference between a

LIMS Cost - Benefit Analysis

COSTS	Year 1	Year 2	Year 3	Year 4	Year 5
Hardware	.....	.....	.....	.....	.....
Software	.....	.....	.....	.....	.....
Services	.....	.....	.....	.....	.....
Staff	.....	.....	.....	.....	.....
Maintenance Contracts	.....	.....	.....	.....	.....
Total Costs	.....	.....	.....	.....	.....
BENEFITS					
Cost Reduction	.....	.....	.....	.....	.....
Cost Avoidance	.....	.....	.....	.....	.....
Revenues	.....	.....	.....	.....	.....
Total Benefits	.....	.....	.....	.....	.....
BENEFITS - COSTS	.....	.....	.....	.....	.....

Figure 12-4 *LIMS financial analysis*

tangible and an intangible benefit is the ability of the organization to assign a value to it. The fact that a benefit cannot be quantified (and therefore intangible) does not lessen its importance to the overall organization.

Financial Analysis

The financial analysis provides an assessment of the proposed system costs *versus* its tangible benefits. Conventions that apply to financial analysis for the justification of systems vary considerably from one organization to another. This section presents the key variables that you should define as a basis for the LIMS cost–benefit analysis.

Figure 12-4 lists the key project-related variables that need to be defined as a starting point for the analysis. They include key elements of the systems costs and its expected benefits along with a projection of when the expected cost or benefit is likely to occur. Each element should be initially stated in today's currency, with an appropriate correction applied to price increases or inflation rates expected in subsequent years.

If the system is purchased most of the hardware and software costs will be incurred during the first year. However, it is possible to defer purchase of selected components, such as terminals or disc storage, until later years when they are needed. Payments may be distributed over several years if the system is leased. In this case, the hardware and software cost may be distributed over the lifetime of the lease. However, interest payments on the lease must be added to the overall cost of the system.

Once these key project-related variables are defined, you should then review the case for the LIMS with the appropriate accounting or financial group from your organization. If necessary, they will add consideration of other variables to your financial case for the LIMS. These include factors specific to your organization and its business function, and any relevant tax codes. These variables include items such as:

- any applicable investment tax credits;
- the amount and rate at which equipment purchase costs can be depreciated;
- the rate that the organization will be taxed on the tangible LIMS benefits;
- the prevailing interest rate that the organization realizes from its cash investments;
- the current rate at which the organization can borrow money.

A discounted cash flow analysis of a LIMS for a laboratory is presented by Golden[2] and critiqued by McDowall.[3] In this example, a LIMS costing $250 000 is installed in a quality control laboratory with a staff of 60. The system is expected to produce productivity savings of 5, 10, and 15% during the first 3 years of operation. Other tangible benefits result in annual savings of $100 000 after 2 years. According to this analysis, the LIMS investment is justified within 2 years.

2 Approvals

The purpose of the approval process is to demonstrate that the project has been well thought out and that there is a reasonable chance of meeting its goals. The expertise of various groups and individuals are considered in a review of the merits and drawbacks of the project. The LIMS project is evaluated relative to other possible uses of the funds to assure that the organization's spending is appropriately targeted.

Types of Approval

Various groups may review the LIMS justification during the approval process. Their role is to verify that all costs have been considered and to ensure that the

[2] J. H. Golden, *Intelligent Instrum. Comput.*, 1985, **3:4**, 13.
[3] R. D. McDowall, in 'Laboratory Information Management Systems – Concepts, Integration and Implementation', ed. R. D. McDowall, Sigma Press, Wilmslow, Cheshire, 1987, p. 53

Type of Review	Groups	Issues/Concerns
FUNCTIONAL	Laboratory Laboratory Clients	• Are the projected benefits realistic? • What will be the impact on the laboratory and those that rely on laboratory services?
TECHNICAL	Information Systems Engineering	• Is the chosen system consistent with the organization's overall automation strategy? • Are the facilities and wiring in place to support the system? • Is the system based on sound and proven technology?
FINANCIAL	Accounting Planning	• What will be the financial impact of the LIMS?
MANAGEMENT	Management	• Does it support strategic goals of the organization? • What are the project management controls? • How will the budget & schedule be monitored?

Figure 12-5 *Groups involved in the LIMS approval process*

project has a reasonable chance of success. Figure 12-5 presents the forms of approval and the various groups that may be involved.

The LIMS Approval Process

The complexity and comprehensiveness of the approval process varies considerably for each organization. The amount of detail needed and the number of approval steps depends on several factors including:

- the cost of the system;
- the number and size of laboratories affected;
- the overall cost of laboratory operations;
- conventions specific to your organization;
- the degree to which key decision makers already support the LIMS.

The following guidelines provide a suggested sequence for obtaining approval for the LIMS. Details regarding each of these activities are discussed in earlier sections of this chapter.

First, complete a survey of LIMS technologies and case histories of past implementations. Focus the study on laboratories that have similar functions and those from similar industries. If possible, arrange for a site visit to targeted laboratories to speak with the principals involved with both the implementation and continued use of the system. Ascertain the potential capabilities, costs, and drawbacks associated with application of the technology to your environment. Determine if it is worth proceeding with the effort involved with the refinement of LIMS requirements and costs.

The next step is to establish implementation and operating costs for the selected LIMS. Prepare a financial and business case with a comparative analysis of the system's costs and benefits. Complete a preliminary review of the LIMS justification with management. Determine if further analysis is required and the exact groups and layers of management that need to be involved with the approval process.

The final step is to proceed with obtaining approvals from the designated groups and individuals. Be sure that your case is well organized and presented in a clear and concise manner.

CHAPTER 13

The LIMS Implementation Infrastructure

The planned LIMS provides new automation technologies that must be established before it is suitable for use by the laboratory. Resources and protocols must be established to install, support, and control these technologies. The magnitude of this infrastructure depends on the complexity, technologies, and risks associated with the chosen implementation approach.

This chapter focuses on the resources, control procedures, and issues relevant to the execution of your LIMS implementation plan.

1 Resources

Hardware and Software

Supplementary hardware and software may be required to support implementation. This includes components in addition to those destined for ultimate use by the laboratory. Their benefits should be carefully balanced against the additional costs and the time needed to gain proficiency in their use. Included are items such as dedicated development systems, software construction aids, and project management tools.

Development Systems

Separate hardware segregates work performed by the LIMS implementation team from that used for normal laboratory operations. A development system is normally a much smaller and cheaper version of the hardware destined for use in production. Its primary objective is to protect production functions and data from potential sources of intrusion and corruption from development. Programs and data are moved to the production system only when they have been properly tested and qualified.

This is an important consideration in cases where the production system supports critical business decisions and functions. It is also important for systems in which high levels of software development are required or anticipated. The development system is dedicated to tasks such as prototyping, coding, testing, training, and software product evaluations. It exists to protect the production system from possible sources of downtime, response time degradation, and data corruption arising from development.

Construction Aids

The efficiency and quality of the LIMS construction can be improved by additional tools. This includes items such as development utilities, code optimizers, test case managers, configuration managers, and Computer Assisted Systems Engineering tools. These tools vary considerably in their exact offerings, costs, and capabilities. They are available from a wide variety of sources.

Project Management Tools

Hardware and software tools exist to assist with the tracking and overall management of the implementation. Project management tools assist with progress tracking, budget monitoring, report issuing, and information sharing. Included are systems for project management, spreadsheets, word processing, and communications.

Personnel

The continued involvement of a project sponsor and a project manager are required for implementation. The sponsor regularly sets aside a small fraction of his or her time to monitor progress and to ensure that the effort remains focused on business goals. The project manager is primarily responsible for completing and coordinating the detailed tasks necessary for the effort. Chapter 10 discusses responsibilities of the project sponsor, project manager, and other resources that may be involved with the implementation. The level of resources required depends heavily on the scope, complexity, technology, and regulatory constraints of the LIMS.

2 Control Procedures

Criteria against which the competence and quality of the implementation will be assessed include its:

- conformance to the planned budget and schedule;
- delivered results and their relevance to business goals.

The success, or lack of success, of the implementation cannot be determined if there is no prior agreement on budgets, schedules, and goals. Maintaining control is difficult, if not impossible, in the absence of clearly defined objectives and parameters to serve as a basis for evaluation. You control the implementation by ensuring that the minimum number of surprises arise during its course. To do so, you need to clearly establish a common understanding of the predicted outcome of the LIMS. This sets the criteria upon which the effectiveness of the entire effort will be evaluated. It also establishes the means for controlling its course.

Without a common understanding, each person and group will assess the effort according to their perception of what the budget, schedule, and benefits should be. Perceptions may be based on information that is incomplete, dated, or from unreliable sources. Problems surface when there are significant disparities between expectations held by those who will ultimately use or pay for the LIMS and those responsible for its installation.

Progress Monitoring

The project plan is an essential tool for maintaining control of the schedule and budget. Chapter 11 details the various elements of a plan, its schedule, budget, and team organization. Progress is monitored by periodically gathering data to answer questions such as:

- what has been done?
- when was it done?
- how much effort was required?
- how much was spent?

At a minimum, the project manager should monitor progress on a weekly basis. He or she compares each period's results with what was expected in the plan. Minor deviations are corrected by altering future work plans or by shifting resources. Corrections may also involve changes to the timetable or budget. Pervasive deviations over a long period indicate that the initial estimates were overly optimistic. In this case, substantial changes to the plan may be warranted.

You should measure progress based primarily on work that has been completed. Each task should have clearly defined end-points and tangible work products. The expected deliverables and quality criteria should be clearly defined. Completion means that the people involved have finished and can be reassigned to other work.

Measuring overall progress based on work in process is more difficult and tends to provide overly optimistic assessments. The progress of in-process work can be judged only if clearly defined metrics are involved. For example, progress on a task called 'convert methods' can be determined if the total number of methods are known. In this case, progress can be based on the percentage of methods already completed. In the absence of metrics, estimating progress based on the proportion of total effort or total elapsed time assumes that the initial estimates are accurate. For a task budgeted for 100 hours, you cannot assume that it is 90% complete because 90 hours has been spent. In reality, the total actual time or effort required is highly variable.

Several standard project management tools are available to help you with monitoring progress.

- A Gantt chart is a series of parallel horizontal bars, with each bar representing a particular task or activity. The *x*-axis of the chart represents dates. The length of each bar indicates the length of time for each task. Tasks that are complete are marked in a special way to distinguish them from those that are not. You can use a Gantt chart to visually depict both the planned and actual status of work elements for a given point in time.

- A Pert chart depicts each activity as a node, with connecting lines between related nodes. Nodes representing completed work are highlighted to separate them from work that is still outstanding. You use a Pert chart to depict the chronological dependencies of work elements and to model how changes in one activity affects others.

A variety of project management software packages can generate Gantt and Pert charts. They are useful for monitoring the implementation's performance *versus* its schedule. However other forms of analysis are required to assess its performance *versus* budget.

Figure 13-1 presents an analysis of tasks completed *versus* the total amount of effort (in hours) spent on implementation. In this case, progress proceeded close to predictions during the first 2 months. Slight deviations were noted in the third month and these increased during the fourth. Progress in terms of the number of tasks completed fell far below plan, to 30% during month 4. This finding is consistent with the fact that the total amount of effort actually spent on implementation is also below plan. The trends are indicative of a project with insufficient resources; either because it is under staffed or because those assigned have been working on other things. Unless corrective action is taken, the implementation will not meet its objectives. In this instance, possible action includes one or more of the following options: extending the completion date, decreasing the scope of work, decreasing work quality levels, or increasing resources.

Progress Reporting and Communications

Good communication is extremely important to the successful implementation of a LIMS. The exact level of detail and information that needs to be communicated varies for each group involved with the implementation.

- Those directly working on the detailed tasks required for implementation should frequently communicate with one another, through both formal and informal channels.
- The project manager should remain aware of the status of each task and the nature of any issues or problem areas. He or she should provide the dialogue necessary to orchestrate the efforts of all groups involved with the implementation effort.
- The project sponsor needs to remain informed about overall progress, key accomplishments, and problem areas, especially those that may impact goals, timing, or funding.

At a minimum, communications with groups outside the core implementation team should occur on a monthly basis. These reports should be short, concise, and clearly provide adequate prior notice of when each group will be affected by the implementation process. Figure 13-2 provides one example of a monthly status report. Reports issued directly to the project manager or project sponsor should provide higher levels of detail.

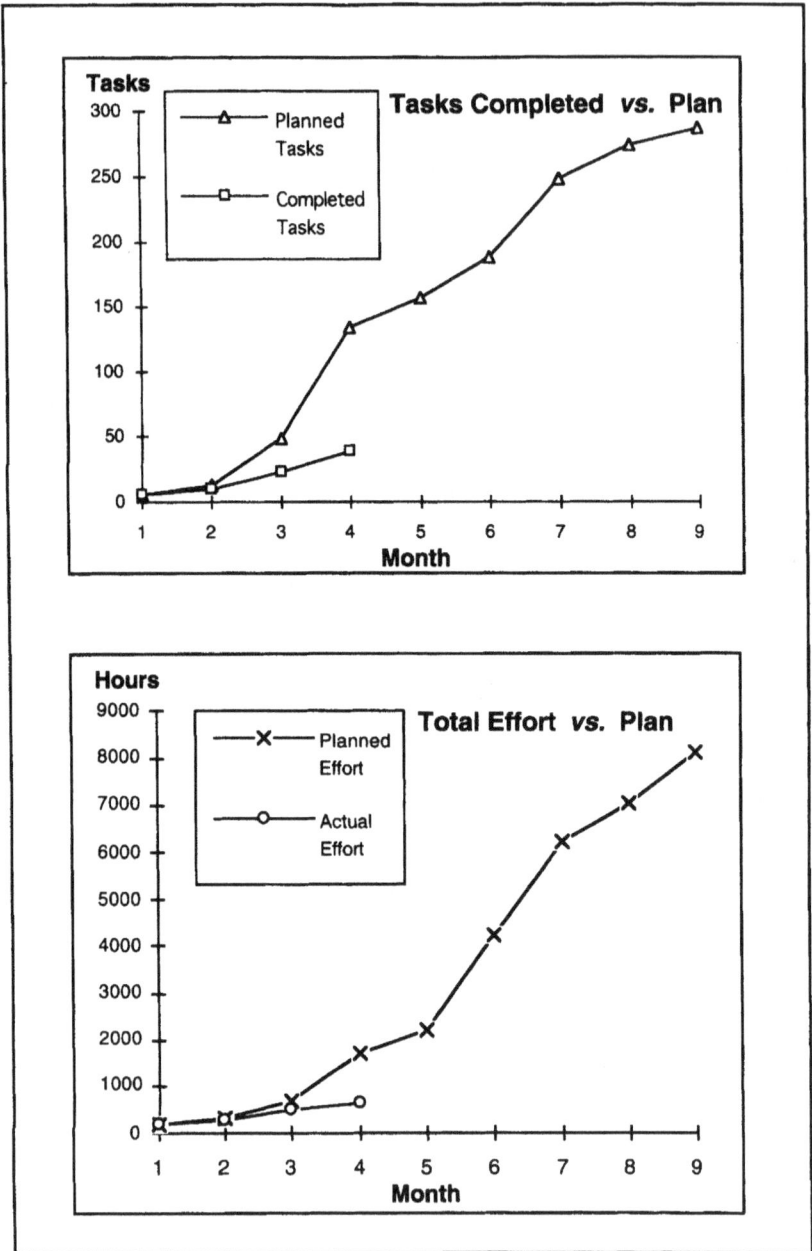

Figure 13-1 *Overall LIMS progress monitoring*

> **Progress Report on LIMS Implementation**
>
> **Accomplishments**
> Tasks completed during this reporting period.
>
> **Non-accomplishments**
> Planned tasks that were not completed during reporting period.
>
> **Problem Areas**
> Problems that have surfaced and what is being done about them.
>
> **Progress**
> Overall progress of project relative to plan.
>
> **Budget**
> Overall use of funds relative to plan.
>
> **Plans**
> Tasks planned for completion during next reporting period.

Figure 13-2 *Example LIMS progress report*

Deliverable Work Product Management

Work products provide tangible evidence that an activity has been completed. They are usually provided in the form of written documentation or a report. Documented work products contain the knowledge gained or agreements reached as a result of an activity. They also certify the completion of key project milestones.

Having work products in the form of documentation has its advantages and disadvantages. The advantages are as follows.

- It provides tangible proof that a given unit of work has been completed.
- The discipline required for writing facilitates presentation of the information in a precise and clear way.
- It provides a tangible means for others to review and substantiate the quality and accuracy of the work effort.
- It serves as a repository of knowledge already acquired. It prevents needless repetition of work that has already been completed.
- They can be presented to auditors to demonstrate compliance.

The disadvantages are as follows.

- Those with high levels of technical skills often dislike documenting what they do. Their personal and professional satisfaction is obtained by solving problems and creating innovative solutions, not in writing down what they did or how they got there.
- Time must be consumed in the writing and production of documentation.
- If the work products are of poor quality or unreadable, the entire documentation effort is wasted.

The scope and complexity of work products varies significantly for each LIMS. They are driven by industry-specific regulations or standard business practices. The exact work products to be produced and their contents should be specified in the project plan (Chapter 11). The following are the key responsibilities of the work product management process.

- *Work product identification and version control.* Each work product should be uniquely identified. With time, each may be revised to reflect new information or changes. Each revision should be assigned a distinct version number and date. In the absence of clear identification, it can be difficult to determine if a particular document contains the most recent information or if it is dated.
- *Work product review and approval.* All delivered work products should be independently reviewed for completeness, accuracy, and readability. An unbiased assessment can only be obtained by individuals or groups who are not directly involved with the actual production of the work product. This review provides useful feedback to its creators who may have either missed or failed to clearly communicate critical elements of the activity. It also ensures that the work product can be understood and used by others.
- *Work product storage and access.* Copies of work products should be available to others. This means that each work product must be stored in a secure location and that processes are in place to ensure that they are accessible to those who need them. Backup copies should also be maintained in case the originals are lost or damaged.

3 Obstacles and Problem Areas

All LIMS implementations will face several obstacles along the way. The best planning, procedures, and staffing cannot totally eliminate all potential problem areas. Many will surface once you are in the midst of implementation. The people involved with the implementation need to be constantly aware of factors that may impede delivery of the LIMS. To do so, they need to rapidly sense, identify, and take appropriate action to diffuse their impact.

The major obstacles to implementation are not generally related to the various elements of the system's hardware and software. They are primarily a result of how the various human and organizational elements are deployed and managed. Technology generally constitutes only about 25% of the solution; the rest relies on people, management, and the organization.

Personnel Issues

People are a critical factor to the success of the LIMS implementation. They represent a diverse range of organizations, disciplines, and interests, some of which are in direct conflict with one another. Chapter 10 presents the various groups and expertise that should be considered for your LIMS effort. The project sponsor and project manager provide critical focus and direction to the overall implementation effort. However, they alone cannot make the LIMS happen. Success depends on their ability to build effective and results-oriented teams. A book by Dyer[1] covers issues and techniques for organization development and team building.

The most common personnel-related obstacles and problem areas are given in the following sections.

Continuity

The involvement of specific members of the project team may be discontinued for a variety of organizational, professional, or personal reasons. They may be transferred to other areas of the organization, find more attractive career opportunities elsewhere, or simply lose interest in the LIMS. The unexpected departure of key elements of the project team, such as the project sponsor or project manager, has a significant effect on the overall effort. Others, who in the past have been significant contributors to progress, will also be missed.

Project schedule alterations may be necessary to accommodate unexpected departures. If replacements are designated, an indoctrination period should be allocated to acquaint them with the background, technology, and procedures associated with the implementation. Most personnel transitions occur suddenly and with little warning. You cannot anticipate the exact changes that will occur, but you can plan for them. The historical staff turnover rate for each group provides a useful approximation for planning purposes.

Work Intensity and Commitment

The LIMS cannot be successful without a team committed to its implementation. Each member of the team contributes specialized knowledge or skills that ultimately influence the quality of the system delivered. Maintaining high levels of work intensity and commitment is difficult, especially when members have other (non-LIMS) obligations to fulfil. To do so, the following should be considered.

- The role and contributions of each individual should be well defined and communicated.
- All efforts result, either directly or indirectly, in a tangible and visible work product. Upon completion, the contribution and achievements of each individual is recognized and acknowledged.

---

[1] W. G. Dyer, 'Team Building Issues and Alternatives', Addison-Wesley, Reading, MA, 1987.

- Communications and meetings are well organized, focused, and convey a respect for individuals' multiple demands and hectic time schedules.

Pride of Ownership

The primary responsibility of the implementation team is to provide a system to serve the organization's needs. In doing so, it is natural for those involved to become professionally and emotionally involved with their work. Laboratory staff and other users of the system need to be intimately involved with the review and approval of critical work products produced by the implementation team. Their role is to correct errors and to suggest changes to improve the utility and quality of the system. It is important to segregate criticism of work products from direct criticism of the project team as individuals and professionals. The project team must be constantly made aware that the ultimate owners of the system will be the organizations who will eventually use and be affected by the LIMS.

Polarization

It is not possible for you to control or influence strongly held beliefs and perceptions within your organization. Each organization has its own set of cultural values, objectives, and styles of management. Managing the implementation requires a high level of skill in working with individuals from highly diverse groups. Individual or group differences should not impede overall communications and teamwork.

Communication

Successful implementation depends on an established network of information exchange and dissemination. It is a critical element of the overall coordination of the implementation effort. With the proper levels of communications:

- people are provided with the necessary information to complete their assignments;
- all affected parties are made aware of schedule changes that affect them directly;
- updates on status, accomplishments, and problem areas are provided on a regular basis;
- issues can be identified, discussed, and resolved in an expedient manner.

Unfortunately, the effort required to maintain proper communications is usually the first casualty during times of tight deadlines and unanticipated problems. In its absence the following occurs.

- Misinformation and rumour develop to fill the voids of missing information.
- Work progress is slowed and schedules slip due to a lack of coordination.
- The quality and timeliness of team efforts decline.

Maintaining the proper level of communications is time consuming and difficult. It requires overcoming barriers imposed by increasing technical specialization, disciplinary boundaries, and specialized vocabularies and acronyms. The challenge grows with the ever-increasing sophistication of both testing and automation technologies. Proper communication is essential to converge the efforts and expertise of increasingly divergent and specialized resources.

Expectations Management

Problems arise when the installed LIMS does not meet the organization's expectations in terms of what it will do and how it is used. The result may be a system that is poorly accepted or totally rejected. The most respected implementations are not the ones with the largest investments or the newest technologies; they are those that truly lived up to their predicted outcomes.

How do you manage the expectations of your organization?

- The expected benefits of the LIMS should be articulated in the goals to be met by the implementation. Chapter 8 discusses potential LIMS goals. The entire organization should be aware of the system's overall goals. Preferably, this will occur at the onset of implementation. Minimally, it needs to occur before it is delivered and installed. The overall development of the system (Chapter 11) needs to maintain its focus on meeting these goals.
- The tangible work products of the implementation effort should be detailed in the project plan, which is discussed in Chapter 11. This includes any related documentation, prototypes, or delivered functions arising during the course of the design and implementation.
- The actual workings of system should be refined as the system evolves, either through a generic Systems Development Life-cycle or through any other development protocol. The organization should be given an opportunity to review, comment on, and request changes to the system being implemented.
- Progress relative to a timetable should be disseminated through periodic reports to the organization. This can occur through a well organized project plan (Chapter 11) as well as periodic progress reports.

Key points to remember in expectation management are to:

- disseminate bad news early. If budgets, schedules, or quality slip be sure that the appropriate parties are provided with advanced notice. With sufficient notice, the organization (and its management) can adapt to unexpected changes. Options are much more limited if unexpected negative results appear suddenly.
- prevent inflated expectations. It is important that you communicate what will happen during implementation. In some cases, it is more important to communicate what will not happen with the system. Within every organization, there exists undocumented, uncommunicated, and frequently unfounded, assumptions regarding what will or will not be provided by an anticipated computer system. The quantity of different perceptions proliferate in direct proportion to the number of people involved. In situations where the perceptions are not consistent with the plans, the plans should be communicated as soon as possible.

Laboratory Staff Participation

The LIMS is a failure if the laboratory cannot or will not use it. If it is used, its quality is determined by how well it fulfils laboratory needs. These needs cannot be established by vendors, programmers, or other technical specialists; they can only be determined by those from the laboratory who will eventually use the system on a daily basis.

Many LIMS are implemented with minimal or no involvement and reviews by the laboratory. The implementation team remains isolated from the individuals who will actually use the system in their daily work. This lack of involvement leads to a poor definition of requirements which carries over into how the system is eventually designed, constructed, and installed.

Project Management

Deficiencies are usually the result of a lack of working experience, either in LIMSs or in project management. Project management deficiencies are usually manifested by difficulties in delegating work, monitoring progress, and maintaining clear communications among the various groups involved. The results are poor planning, poor communications, and schedule slippage. These deficiencies can be overcome by extensive training, mentoring by the project sponsor, and a comprehensive project plan (Chapter 11).

Development Techniques

LIMS implementation projects are often compromised by technical specialists who lack the training or experience of the implementation of complex multi-user automation technologies. The results are *ad hoc* techniques and complex solutions which are inconsistent with laboratory needs. Progress slips as developers are inefficiently spending most of their time uncovering errors and omissions from either poor or non-existent designs.

CHAPTER 14

Managing Change

The LIMS inevitably introduces changes in the way people work and interact with one another. Both the laboratory and the enterprise must be prepared to accommodate changes brought on by the new system. Unanticipated changes also occur during the course of implementation, usually at the most inopportune moments. These may be a result of unexpected shifts in the organization's management, personnel, business climate, or regulations. More often, it is a result of new, previously unanticipated, information which can alter the system's design or its implementation. An implementation that fails to accommodate new facts and new needs can miss opportunities to improve the system's overall quality and effectiveness. However, one that has too many changes may lose its focus and never be completed.

Chapter 6 discusses the impact of the LIMS on both individuals working in the laboratory and the organization itself. This chapter discusses the nature of the changes and the process of managing them.

1 The Nature of Change

With the passage of time, change is inevitable. The longer the time period, the greater the change. In many modern societies, the pace of change has increased substantially over the past three generations. Before that, people were born, raised, made their living, bore offspring, and died within a few miles of their ancestor's birth place. Technology advances in areas such as transportation, communications, and automation have significantly accelerated the pace of change. Today, it is not unusual to have generations of a family residing in physically separate cities or countries. In some instances, core family units are also geographically scattered.

The same is true for our businesses and other organizations. Today's enterprises are increasingly geographically dispersed, less hierarchical, and subject to substantial, rapid, and unanticipated changes in their organization, operations, and relationships with customers. This has occurred in response to ever-changing business demands, increasingly assertive regulators, and expanding customer needs. The organization's survival will be dictated by its ability to respond to unanticipated changes. Most of these changes would not be possible without the enabling technologies that allow us to be more efficient, faster, and provide higher levels of quality.

Laboratory capabilities have significantly increased over the past few decades. We can now analyse many more components at much lower levels of detection because of innovations in ever-more sophisticated instrumentation and novel testing techniques. It is possible for us to obtain a great deal of information from extremely small amounts of sample. Today, many recently trained scientists are totally unfamiliar with most of the test methods, apparatus, and reagents commonly used 20 or 30 years ago. It is likely that today's testing technologies will similarly be dated by scientific advances over the next decade or two.

Rapid changes can also be expected in the way laboratories operate. They must respond rapidly to unanticipated work volume shifts, newer testing technologies, increasingly automated instrumentation, and changing management policies.

Although change is inevitable, many people and organizations resist it. This is a natural initial reaction, especially for those comfortable with the *status quo*. The new way of working may be only remotely related to the past way of doing things. Many may see any proposed change as a potential threat to their security, status, and ability to do their work. Initially, resistance is generally caused by a lack of knowledge about the actual effects that the change will have, both to the individuals themselves and to their organizations. If this void is not filled with concrete and credible data, then it will be filled by rumour and misinformation. The rumours are usually much worse and more negative than the reality.

Both Handy[1] and Senge[2] equate the process of change with learning. Handy states, 'Those who are always learning are those who can ride the waves of change and who see a changing world as full of opportunities, rather than damages. They are the ones most likely to be the survivors in a time of discontinuity.' Senge presents the concept of a learning organization which is 'an organization that is continually expanding its capacity to create its future. For such an organization, it is not enough merely to survive.' Both authors discuss factors that lubricate as well as impede change.

2 Types of Change

Change impacts a LIMS in two ways. First, introduction of the LIMS may drastically change the way that the laboratory and other affected organizations operate. Conversely, groups outside the direct control of the project team may impose changes to the implementation effort. The LIMS implementation effort is both an agent and an object of change. It introduces changes to the laboratory and, at the same time, is subject to changes demanded by others.

The LIMS as an Agent of Change

Changes accompanying the LIMS are not limited to the newly introduced hardware and software technologies. To be effective, it requires a new set of skills and a different way of operation within the laboratory. The LIMS may replace an

---

[1] C. Handy, 'The Age of Unreason', Harvard Business School Press, Boston, MA, 1989.
[2] P. M. Senge, 'The Fifth Disipline: the Art and Practice of the Learning Organization', Doubleday/Currency, New York, 1990.

existing manual or automated system. However, if it is to be used as a source of improvement, it should not exactly emulate the system or systems it replaces. If it does, then it merely inherits and proliferates the weaknesses and limitations of the past.

A sound programme of communications, education, and training can greatly facilitate the acceptance of forthcoming changes to the laboratory. This should, in some way, include all individuals and groups targeted to use the system. Such a programme should be instituted as early as possible, not just prior to turnover of the system to the laboratory. It should clearly articulate what is known regarding the new skills required and how jobs will be changed. What is not known or is undefined at a particular point in time should also be communicated.

The LIMS as an Object of Change

The LIMS itself will also be subject to changes. As implementation proceeds, new information and new business needs will emerge. In a sense, 'Developing a LIMS is like changing the tyres on your car while its still rolling'.[3] The organization targeted for the LIMS may be in a state of constant evolution and change in response to new technical, business, and customer mandates. These changes may affect both the scope and complexity of the LIMS. To be successful, the implementation effort should minimally have a way of considering and, if warranted, accommodating the changes.

A project that encounters numerous requests for new functions or changes may, in fact, be one in which the proper data was not adequately gathered during the early stages. This commonly occurs when the needs assessment process (Chapter 9) fails to involve the right people or is incomplete. New requirements should be those that were not known to the laboratory during the early stages. They surface once implementation has started. Incomplete requirements are those that were known, but which were not uncovered by the project team; either because they asked the wrong people or because they failed to ask the right questions.

3 Why Manage Change?

To some extent, the refinement of information as the project progresses is inevitable. With the passage of time, new business needs emerge and old information is refined or corrected. The overall design of the LIMS may rely on assumptions based on previously gathered data that may be incomplete or incorrect. A process is needed to accommodate new information and viewpoints and, if necessary, make any warranted changes. Otherwise the delivered system may not adequately serve the organization for which it is intended. A LIMS that meets its published requirements will not be successful if the document upon which it is based fails to reflect the needs of the organization. At the same time, it may not be practical to address all requested changes. The implementation's

---

[3] K. E. Blick, 'Computerization of Clinical Laboratories: Past, Present, and Future', Fourth International LIMS Conference, Pittsburgh, PA, 1990.

success can be seriously compromised if the magnitude of the changes exceeds the projects available resources and timetable.

A change-management process is needed:

- to provide forums that facilitate the exchange of information between those charged with implementation and those who will ultimately own and use the system;
- to ensure that mechanisms exist to identify and remedy cases in which the implementation is proceeding based on incorrect or incomplete information;
- to provide a way of identifying, reviewing, and accounting for additional resources needed and time required;
- to provide a chronological record of revisions to previously published and approved project-related documents.

4 The Change Management Process

It is not possible for you to control change but it is possible for you to manage it. A sound change management programme incorporates procedures, responsibilities, and forums to deal with resistance to the new system. It also provides mechanisms to respond to new information or new business needs. The LIMS changes the organization and the people who work in it. And the organization itself imparts demands on and change to how the LIMS is implemented. On a case by case basis, one of the most pressing issues becomes, 'Who should change; the LIMS, the organization, or both?'

A sound change management process ensures that the efforts of those involved with delivery and implementation of the system are congruent with those who must use it. It is difficult to manage this process unless you have a sound base of understanding of the organization's needs (Chapter 9) and operations (Chapters 3–5). Change management involves changing the beliefs, behaviours, and skills of two groups: the laboratory, with other groups who will inherit the LIMS, and the core project team charged with its implementation.

For the laboratory users to embrace the changes brought about by the LIMS, they need to feel that the system is worthwhile; that it provides a net improvement, either to themselves personally or to the organization as a whole. This involves an awareness of the system's objectives and goals early on. It also requires constant dialogue between the project team and users. Too frequently, the project team work long hours in isolation. They do not take the time to talk or communicate with anyone else. Maintaining dialogue with the laboratory or anyone else is a luxury they can ill afford. In a way they are sending a hidden message. By doing so, they are telling the laboratory, 'We're busy trying to figure out how you will work in the future and we'll tell you about it when we are ready.' By doing so, they prevent the people who are expected to inherit the system from actively participating in its evolution and design. When it comes to the time to deliver the system, the users feel alienated and devalued because their concerns and feelings have never been solicited or considered. The result may be outright resistance or subtle efforts to undermine the system.

In a classic 1954 article, 'How to deal with resistance to change',[4] Paul R. Lawrence discusses the importance of considering the social impact of technology changes on organizations. Often too much emphasis is placed on the technology elements, and there is a failure to consider the importance of the existing, but informally established, mechanisms of interactions, information exchange, and problem solving. Lawrence claims that each organization informally establishes change management mechanisms. Too often, the implementation of technology changes fails to consider and capitalize on existing ways that the organization itself adapts to change.

'We know that people who are working closely with one another continually swap ideas about short cuts and minor changes in procedure that are adopted so easily and naturally that we seldom notice them or even think of them as change. The point is that because these people work so closely with one another, they intuitively understand and take account of the existing social arrangements for work and so feel no threat to themselves in such everyday changes.

By contrast, management actions leading to what we commonly label 'change' are usually initiated outside the small work group by staff people. These are the changes that we noticed and the ones that most frequently bring on symptoms of resistance. By the very nature of their work, most of our staff specialists in industry do not have the intimate contact with operating groups that allows them to acquire an intuitive understanding of the complex social arrangements which their ideas may affect. Neither do our staff specialists always have the day-to-day dealings with operating people that lead them to develop a natural respect for the knowledge and skill of these people. As a result, all too often the men behave in a way that threatens and disrupts the established social relationships. And the tragedy is that so many of these upsets are inadvertent and unnecessary.'

Change management is not the same as project management. Project managers supervise the execution and completion of specific tasks, with deadlines, deliverable work products, and available resources. Change management affects the dynamics of an established project plan. It does so by the addition, modification, or deletion of tasks and deadlines. Change management involves an holistic view of the system; including the technical components, its acceptance by users, and its perceived effectiveness within the organization. It considers if activities should be undertaken in the first place, if additional work is warranted, or if the project plan should be modified. Project management defines what it takes to change the tyre on an automobile. Change management establishes how this is done while the car is moving ahead at full speed.

An established change control process is important. It should be reviewed, approved, and understood by everyone involved with both the delivery and use of the LIMS. It should establish an environment conducive to open discussions and exchange of views. The change control process provides a means of soliciting the ideas, solutions, and perspectives of others. Any input that may ultimately improve the quality of the system should be encouraged. It should all be

[4] P. R. Lawrence, *Harvard Bus. Rev.*, 1954, **32:3** 49.

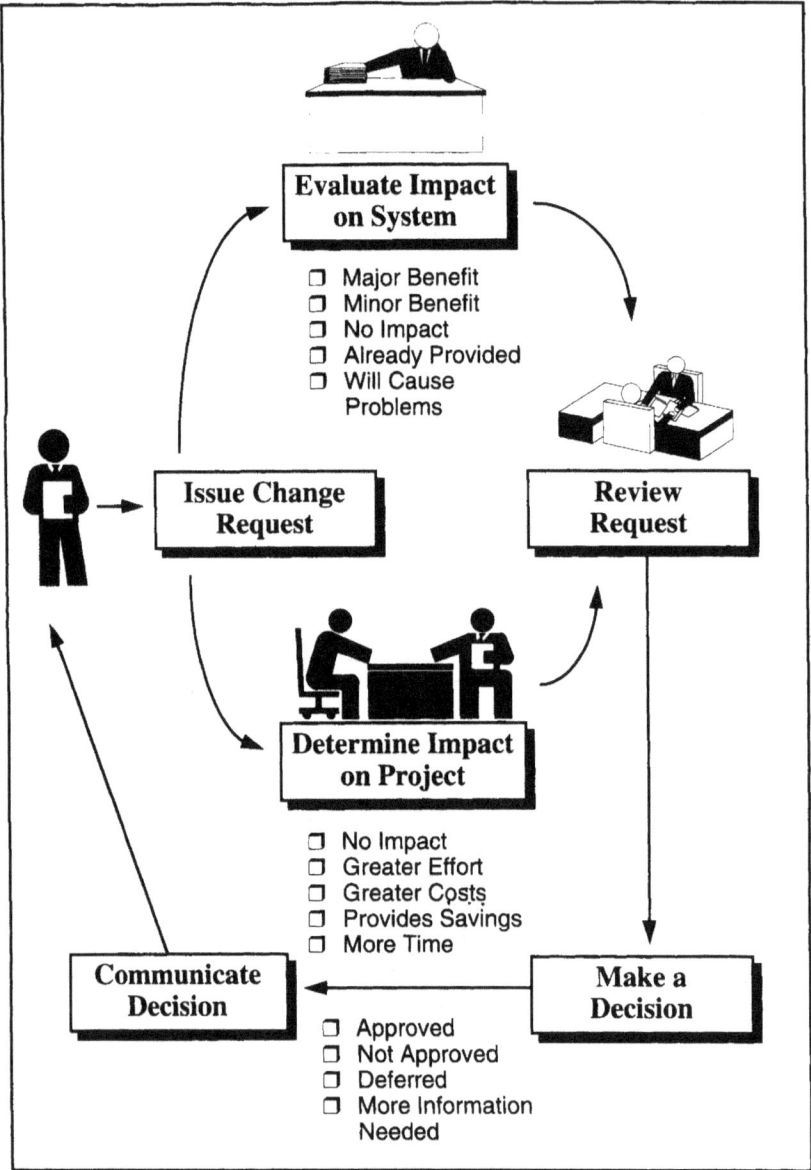

Figure 14-1 *The change control process*

considered carefully, even though it may potentially alter the existing course of implementation. A typical change control process is shown is Figure 14-1.

Each change control process includes the following elements.

- *Responsibilities*. Someone must be designated as a coordinator of the change management process. His or her principal responsibility is to ensure that change requests are processed expediently and that the proper people are involved with the review and decision-making process. The project sponsor must be involved with changes that affect the overall costs, timetable, and benefits of the system.

- *Initiation.* A change can be requested by anyone. Basic information regarding the requested change is obtained. This includes the nature of the requested change, why it is necessary, and the person or persons requesting the change. Following a cursory review, the requester is contacted if further information is needed to fully understand the request.
- *Impact analysis.* The requested change is evaluated for its impact in terms of time, cost, resources, and benefits. This involves its effect on the organization as well as its effect on the implementation. The appropriate decision makers are identified.
- *Review and decision.* Results of the impact analysis are presented and reviewed with the decision makers. They consider the impact of the change and make a decision. Possible actions are to accept, reject, or defer the request.
- *Feedback.* The original requesters are informed of the impact and decision made on their request.
- *Logging.* Information regarding each requested change, its impact, and its disposition is recorded. For approved changes, this provides an historical record of the system's evolution and who approved each change. During the course of a project the same changes may be requested by different parties. The record details how previous similar requests were handled.

CHAPTER 15

Post-implementation Considerations

Once established, the LIMS provides the laboratory with enhanced information management capabilities. The original implementation effort concludes when the delivered system meets its targeted goals and needs. Chapters 8 and 9 discuss LIMS goals and needs assessments. For systems that are large in scope, implementation should be segregated into clearly defined building blocks, each meeting a subset of the overall system's goals. Such a strategy should be inherent in the LIMS implementation plan as discussed in Chapter 11.

Portions of the LIMS that have been completed represent an investment by the organization. The investment can be quantified by the funds and effort expended to date. The LIMS can also represent a valuable resource to the organization. It does so by providing the organization with new capabilities and operating efficiencies. One of the most important post-implementation considerations is to safeguard the organization's past investment in this newly established information management resource. To do so, the organization needs to establish a post-implementation support infrastructure consisting of the appropriate personnel, tools, funding, and procedures. This infrastructure ensures that the system continues to service laboratory or other business needs in a reliable and predictable manner. We need to support the LIMS as we support the various machines that we often take for granted in our daily lives. Without the proper fuel, lubrication, fluids, maintenance, and monitoring, their reliability deteriorates and they eventually cease to function altogether.

Once implemented, the LIMS does not become a static entity. In fact, it continues to develop further. A typical evolutionary sequence is depicted in Figure 15-1. Initially, the organization seeks to expand its utilization in new and innovative ways. System enhancements emerge in response to new business demands and work volume growth. The new and enhanced capabilities are implemented at the same time that established functions are maintained. After a period of time, the technical foundation of the system becomes obsolete. The effort required to maintain and enhance the system becomes increasingly burdensome. The system undergoes a period in which it may be technically dated, but still remain functionally useful to the organization. Eventually, the costs and constraints of the older technologies drive the organization to consider either a total or partial replacement of the LIMS.

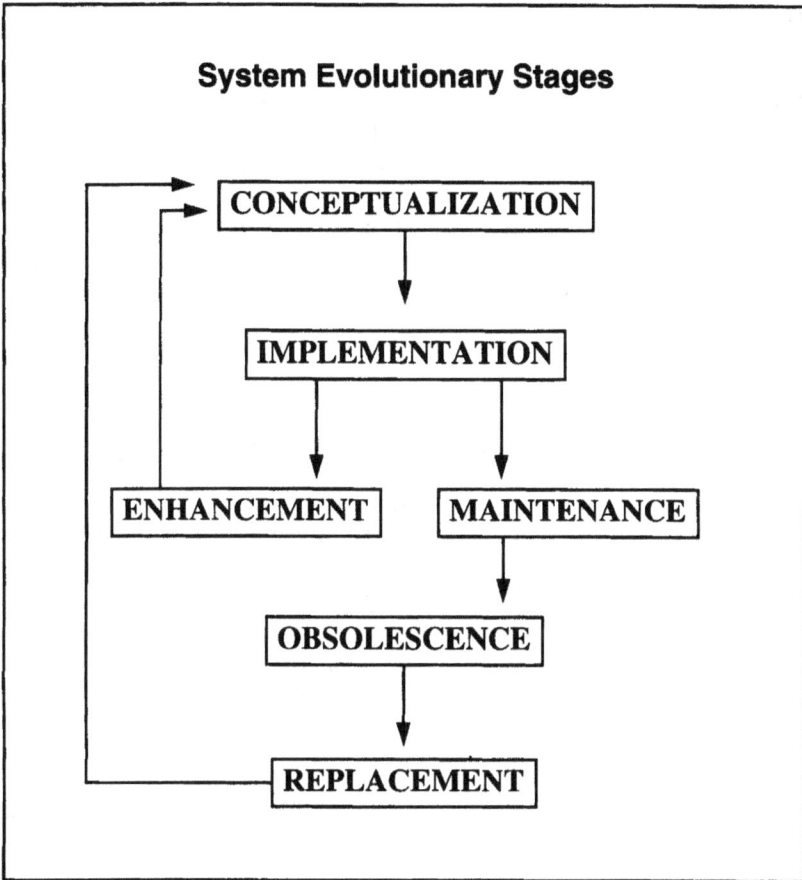

Figure 15-1 *Stages in the evolution of a LIMS system*

This chapter discusses two post-implementation considerations for a LIMS: its support infrastructure and its evolutionary stages.

1 Post-implementation Support Infrastructure

Several elements make up the LIMS support infrastructure. They collectively support the organization in its use of the system and ensure that the systems hardware and software continue to operate in a reliable manner. This section describes the resources and procedures that comprise the support infrastructure.

Resources

Resources required to maintain the LIMS include personnel to support the system, funding for expendable supplies, and services in the form of both hardware and software maintenance agreements.

Personnel

The various roles needed for post-implementation support of your LIMS are presented in the following sections. Each role consists of a set of related responsibilities. Each may be filled by one or more persons, depending on the size and complexity of the installed system. In some instances, a small team or even a single individual may be held responsible for all aspects of maintaining the system. The exact individuals assigned may come from several sources. They may be from the information systems group, the laboratory, other support groups within the organization, or from outside.

LIMS Administration. Someone should serve as the focal point of communication and coordination of LIMS-related services with laboratory users, management, vendors, contractors, and all other groups that either use or play a role in continued operations of the installed LIMS. This includes responsibilities such as:

- maintaining and monitoring the LIMS support budget;
- supervising the efforts of other support staff, vendors, and contractors;
- coordinating the dissemination of information to management or others to expedite the analysis and decision making on LIMS-related issues;
- coordinating the change requests and other established administrative processes;
- ensuring that all components of the LIMS support infrastructure operates in a consistent manner with standards established by the organization and regulatory agencies.

Systems Management. Someone should be primarily responsible for overall maintenance and continued reliable operation and availability of the system. This includes all components of its hardware, software, and networking. Typical responsibilities are to:

- ensure that routine and remedial maintenance activities are performed in a timely manner by appropriately qualified personnel;
- ensure reliable operating performance and availability of the system for the laboratory and other users;
- define and document standard operating procedures for routinely scheduled maintenance activities such as back-ups, file purging, and archiving;
- ensure that operators, vendors, or others comply with previously defined procedures and maintenance schedules;
- ensure timely restoration of the system in the event of unexpected failures.

System Operations. Individuals need to perform periodic maintenance tasks according to standard operating procedures approved by the systems manager. These include routine systems monitoring and scheduled maintenance of the LIMS hardware and software elements. Unexpected problems are brought to the attention of systems management.

Specific duties related to system operations include:

- providing regular backups of program files and the database;

- maintaining the disc storage areas which includes periodic compression of fragmented areas and archiving;
- monitoring and periodic maintenance of the database management system;
- monitoring the various system, software, and error logs created by the software;
- monitoring the system's performance and resource utilization.

Development. Developers correct errors and solve problems with currently installed software. They also add new functions and capabilities to the system. Problem solving is mandatory to ensure that the LIMS operates, and will continue to operate, in a reliable manner. Many problems will not surface until the system has been in use for some time. Enhancements are discretionary. They serve to provide new functions and capabilities beyond those covered by the initial needs targeted for the LIMS. The need for and approval of enhancements should be carefully considered. Each should add value to the overall operation of the organization.

Maintenance Agreements

Maintenance contracts are available from vendors. These agreements provide you with the expertise, personnel, and upgrades to ensure that the technical hardware and software elements operate in a reliable manner. These agreements include elements specific to vendor products. They generally do not include items specific to your system. However, the exact elements covered by maintenance agreements vary significantly and may, in some cases, be negotiated.

Hardware Maintenance Agreements. For a defined annual price, hardware maintenance agreements provide services and replacement parts to maximize working times and reliable operations of the LIMS.

Contracts are normally issued following a site inspection. At this time, the vendor checks the initial condition of the hardware to ensure that it was installed properly. They also check for environmental conditions that contribute to frequent hardware failures such as inadequate temperature controls, poor electrical power quality, and the presence of dust or corrosive vapours.

From a budgetary perspective, the principal advantage of these agreements is that they provide predictability in hardware maintenance costs. A single repair can easily cost $10–100 000. For many organizations, it is difficult to obtain unbudgeted funding of this magnitude. Maintenance contracts allow you to distribute these costs over several years.

The costs of hardware maintenance agreements should not be compared solely with the (unknown) costs of discretely paying for parts and services as needed; it should be balanced against the larger costs that LIMS downtime may have on your organization.

Generally, hardware maintenance agreements provide the following services.
- *Periodic testing.* Key hardware components such as the processor and disc drive units are regularly tested. Diagnostic programs are run to ensure that principal components are working properly.

- *Regular preventative maintenance*. This may include cleaning, parts replacement, or filter changes on key hardware elements.
- *Hardware updates*. Hardware vendors constantly engineer newer and more reliable parts for their systems. Parts that break down frequently are generally replaced or re-engineered with newer versions. Hardware maintenance contracts may include these updates.
- *Diagnostic telephone support*. The vendor provides a designated contact for telephone support. When problems arise, requests for assistance are directed to the designated contact. The vendor determines the nature of the problem and implements corrective measures. If necessary, a qualified service engineer is dispatched to implement the corrective action.
- *Replacement parts*. Hardware components that break during the normal course of operation are replaced by the vendor.

The exact costs of hardware maintenance contracts vary considerably. They are directly related to the supplier's costs of providing services. Contributing factors include the initial cost of the hardware, the age of the equipment, the hours of service coverage, and the elapsed time between placement of a service call and the on-site arrival of a service engineer.

- *Initial costs of hardware*. Annual costs are normally 10–15% of the initial hardware purchase price. The exact percentage varies according to the level of services provided and the age of the system.
- *Coverage hours*. Costs increase for contracts that provide increased or special coverage. Normal coverage is defined relative to normal working hours in the time zone where the service organization is geographically located.
- *Service response cycle time*. Costs increase for contracts that provide faster response to service calls. Normal response is normally defined in terms of working hours or days from the time that assistance is initially requested. If faster response is desired, the provider may hire additional personnel and maintain an inventory of spare parts dedicated to your site. These expenses are reflected in higher maintenance contract costs.
- *Age of hardware*. Contract prices increases as the hardware ages and becomes more prone to failure. Service costs for hardware that is no longer manufactured rise significantly, mainly because replacement parts become increasingly difficult to obtain.

Software Maintenance Agreements. Services to support continued and reliable operation of various software packages on your system are provided by the respective vendors of each package. This includes the computers operating system, database, utilities, and applications. Separate agreements are normally required with each software vendor.

The nature of the support provided by software maintenance contracts varies considerably. Services typically provided by these agreements include:

- *Technical support*. Telephone support is provided for questions regarding the software's use and configuration. This provides useful assistance with problem diagnosis and exploring more sophisticated uses of the software.

- *Software revisions*. Periodic software revisions and enhancements are issued to solve identified problems and to provide functional and performance enhancements.
- *Technical communications*. Periodic technical reports and newsletters are issued. They document future directions of the software, known problem areas, and how other customers have applied the software to their environment.
- *Software upgrades*. The vendor may provide customers with service agreements with discounts on other new or improved software packages.

Pricing policies for software maintenance agreements vary from one vendor to another. They are periodically revised in response to customer needs. Factors that affect the software support costs include the initial price of the software, the number of users, the size of the computer processor, and the specific services provided.

The principal advantage of software support agreements is the access provided to the vendor's technical experts. These individuals are fluent in the operation, configuration, and unique features of their software. They can enhance the capabilities of your LIMS by providing your project team with the benefit of their accumulated knowledge and experience with the software and its implementation in a variety of laboratories.

Procedures

Systems Operations

Several procedures are required for the routine and proactive monitoring, maintenance, and operations of the system. The procedures should be documented. Each time a procedure is carried out, the dates and names of responsible individuals should be recorded. Events that are unexpected or which require further attention should be recorded and the system manager notified.

Periodic Monitoring. Someone should be designated to frequently monitor key operating variables of the working system. The monitored variables should have pre-defined acceptable outcomes. Excursions from the expected ranges should be investigated and corrections implemented. This includes monitoring the:

- storage capacity and utilization of the discs and database;
- error or warning messages from the system;
- system performance and response time.

Preventative Maintenance. Regular procedures should be established to frequently perform activities to prevent problems and loss of reliable operations. The exact procedures required are specific to the hardware and software components of your system and its operating loads, but can include activities such as:

- back-up of data and programs;
- reduction of file fragmentation on discs;
- cleaning of printers and other parts;
- database reorganization and maintenance.

Routine Operations. Several procedures are necessary to ensure adequate routine daily operations of the system, such as:

- ensuring that printers have adequate paper supplies;
- if necessary, loading and running scheduled programs;
- distribution of printed reports to the appropriate individuals or groups.

LIMS Document and Records Control. Procedures and responsibilities should be defined for the creation, storage, indexing of, and access to LIMS-related documents, procedures, manuals, and records. The procedures should ensure that the correct versions of documents are maintained and segregated from versions that are out of date. They should also facilitate access to information needed to resolve problems with the system or for audits.

Remedial Maintenance

Remedial maintenance activities are undertaken in response to unexpected failures or problems with the LIMS. Unlike preventative maintenance, which is proactive, remedial maintenance is reactive. The responsibilities and general procedures for the expedient identification, coordination, and resolution of system problems should be well understood and communicated to everyone involved with the use and support of the system.

Support for System Users

Procedures and responsibilities should be established to assist those who use the system and its programs. This involves answering questions, providing training, and distributing user manuals. The exact level of assistance needed diminishes as users gain experience of the system. Periodically, with staff turnover, new users are established on the LIMS as former users are removed from the system.

Problem Resolution Procedures

Procedures should be in place for the identification, recording, investigation, and resolution of all problems with the system. Problems may take the form of hardware or software failures, unclear manuals, undefined responsibilities, or undesirable behaviour of the system. Not all problems identified may be immediately resolvable. However, a written record of the incident and its observed behaviour may facilitate its resolution in the future.

Configuration Management

A way of recording and tracking changes to the system's hardware and software configuration should be established. This involves any additions or modifications of the system's hardware configuration or software programs, either supplied by vendors or internally developed. Procedures should be in place to ensure that the change does not directly or indirectly affect the reliability of other components of the system. Records should also be maintained to trace when each change was implemented.

Disaster Recovery Planning

A disaster recovery plan includes detailed steps necessary to restore and continue operations of the LIMS in the event of key component failures. This includes the loss of key hardware elements such as the processor or storage devices. It may even involve the total loss of all hardware due to fire, flood, sabotage, or other catastrophic events. Elements of a disaster recovery plan include:

- an assessment of the system's criticality and the effect that its loss will have on the organization. The costs and extent of measures should be consistent with the value that the organization places on continued operation of the LIMS.
- periodic back-up of the system's programs and data for restoration of the system. At least one copy should be stored separately from the key hardware components, preferably in a separate building.
- provisions for the expedient acquisition or use of replacement hardware.
- designation of individuals responsible for each element of the plan.
- training in plan execution.
- procedures for the continuance of laboratory services while the LIMS is unavailable.

2 Evolution of the LIMS

Systems Growth and Expansion

The initial LIMS implementation delivers information management capabilities defined by the project's original scope. This milestone generally marks the beginning, not the end, of the organization's exploitation of automation technology. Much of its potential is not fully appreciated until the organization has worked with the LIMS for some time. The users and the organization becomes increasingly aware of both the strengths and constraints of the system as it establishes itself into the routine of daily laboratory operations.

The system grows as new and creative ways of using the LIMS emerge. It also grows because of expanding demands from increased laboratory workloads and user demands.

Expansion of Requirements

The scope of the initial LIMS implementation does not usually address all requirements. Some are intentionally deferred from the initial phase of implementation due to constraints in the technology or in available resources. Others surface after the organization has used the system for some time. In addressing new requirements, attention must be given to a careful definition of needs to be served (Chapter 9), planning elements (Chapter 10), and justification (Chapter 12) of proposed LIMS enhancements.

Expansion of Capacity

To adequately serve the organization, the LIMS must expand consistently with the growth of the organization's work volume. This may mean the addition of new hardware to accommodate increased storage needs and more users. As demands on the system increase, system upgrades may be needed to improve computer throughput, response time, and communications capabilities. The need to expand capacity can be determined by monitoring trends in the system's resource utilization and performance.

LIMS System Obsolescence

The LIMS will undergo two stages of obsolescence: technical and functional. Technical obsolescence occurs first. It happens when the underlying hardware and software become dated. In the marketplace, they are replaced by components that are faster, more efficient, and cheaper, or which possess higher levels of functionality. Innovations in hardware and software occur so rapidly that each offering become technically obsolete within a very short time. This does not mean that you should constantly replace your system with the most recent product offerings. Given the appropriate levels of resources and maintenance, products that are technically dated can continue to reliably fulfil the organization's needs for many years.

Functional obsolescence occurs when continued reliability of the LIMS deteriorates as it becomes increasingly costly and difficult to operate. The hardware and software components may no longer be marketed and, with the passage of time, vendor support may become less reliable. Problems become increasingly frequent and they take much longer to resolve. As the customer base declines, so does the vendor's internal expertise and the quality of support.

Functional obsolescence also occurs when design constraints of the system cannot accommodate new business demands. This may be a result of new workloads, security requirements, performance demands, or organization-wide standards that were not identified when the original system was designed.

The Next-generation LIMS

Planning of the replacement next-generation LIMS should start immediately after the initial phases of implementation has been completed. If care is taken in the assessment of the organization's needs, if careful thought is given to its selection, and if the implementation is properly managed, a LIMS should last from 8 to 12 years before a replacement is warranted.

You should capitalize on the lessons learned from the implementation and operation of the current system. Both its strengths and weaknesses should be continuously monitored and assessed. Problems encountered with its implementation and maintenance should also be noted. Lessons learned from the current LIMS should be applied to the needs assessment, selection, and implementation of its replacement.

It is important for the organization to maintain an awareness of developments and trends in LIMS technologies. Frequently, the people involved concentrate their total attention on the installation, maintenance, and use of the current system. Some time should be spent keeping track of market developments in LIMS-related products. Changes in this area occur rapidly and many seemingly promising solutions fail to realize commercial acceptance. Each innovation should be critically evaluated for its potential applicability to the next-generation LIMS. It is also important to examine the experiences of other organizations in the implementation and use of their LIMS.

The transition from the current to the replacement LIMS requires careful consideration. Issues regarding the movement of data and procedures need to be resolved prior to the conversion. Will historical data and procedures from the old LIMS be transferred to the new system? If so, how will it be transferred and converted? How will the laboratory transfer work in progress from one system to the other? When will the old system be discontinued?

The next-generation system extends the information management tools and capabilities made possible by the LIMS. Its only limit is the vision, creativity, and discipline of the people and the organization charged with its implementation and its management.

Subject index

www.ingramcontent.com/pod-product-compliance
Lightning Source LLC
Chambersburg PA
CBHW070719220326
41598CB00024BA/3226